Global Environmental Governance in the Information Age

This book examines the impact of current and emerging digital technologies on global environmental governance, and in particular on environmental civil society organizations.

Technological innovations are constantly emerging: internet, social media platforms, blockchains, big data, and artificial intelligence are some of the most common or promising digital technologies of our times. Through case studies and the analysis of concrete applications of digital technologies, this book shows how these digital technologies can be deployed to support global environmental governance, and in particular a multi-stakeholder approach to the protection of the environment. It provides an overview of the diverse uses of these digital technologies by civil society organizations (CSOs) in global environmental governance. In this fast-changing context, the capacity of environmental CSOs to employ digital technologies, can strengthen their participation in global environmental governance. Their key roles, including advocacy, monitoring, knowledge production, fundraising, nudging individual behaviors, and project implementation, greatly benefit from the use of these technologies. By examining some of the most-utilized current digital technologies and presenting some of the most prominent emerging ones, this book aims to illustrate how digital technologies can trigger some evolutions in global environmental governance.

This book will appeal to scholars and students of environmental studies, political science, international relations, geography, and communication studies along with policy makers and communication specialists from the environmental community.

Jérôme Duberry is Post-Doc Researcher and Lecturer at the Centre de Compétences Dusan Sidjanski en Etudes Européennes, Global Studies Institute, University of Geneva, and Lecturer and Research Associate at the Graduate Institute of International and Development Studies. His research activities revolve around the contemporary transformations of society and international institutions in a process of adaptation to the emergence of new technologies. He is particularly interested in innovative public and private governance systems, and examines how information and communication technologies (ICTs) influence global multi-stakeholder decision-making processes. He holds a PhD in International Relations from the Pompeu Fabra University, Barcelona.

Emerging Technologies, Ethics and International Affairs
Series Editors
Steven Barela, Jai C. Galliott, Avery Plaw, Katina Michael

This series examines the crucial ethical, legal and public policy questions arising from or exacerbated by the design, development and eventual adoption of new technologies across all related fields, from education and engineering to medicine and military affairs. The books revolve around two key themes:

- Moral issues in research, engineering and design
- Ethical, legal and political/policy issues in the use and regulation of technology

This series encourages submission of cutting-edge research monographs and edited collections with a particular focus on forward-looking ideas concerning innovative or as yet undeveloped technologies. Whilst there is an expectation that authors will be well grounded in philosophy, law or political science, consideration will be given to future-orientated works that cross these disciplinary boundaries. The interdisciplinary nature of the series editorial team offers the best possible examination of works that address the 'ethical, legal and social' implications of emerging technologies.

Cyber Attacks and International Law on the Use of Force
The Turn to Information Ethics
Samuli Haataja

Global Environmental Governance in the Information Age
Civil Society Organizations and Digital Media
Jérôme Duberry

Technology and Agency in International Relations
Edited by Marijn Hoijtink and Matthias Leese

For more information about this series, please visit: www.routledge.com/ Emerging-Technologies-Ethics-and-International-Affairs/book-series/ASHSER-1408

Global Environmental Governance in the Information Age

Civil Society Organizations
and Digital Media

Jérôme Duberry

Routledge
Taylor & Francis Group

LONDON AND NEW YORK

First published 2019
by Routledge
4 Park Square, Milton Park, Abingdon, Oxon OX14 4RN

and by Routledge
605 Third Avenue, New York, NY 10017

First issued in paperback 2021

Routledge is an imprint of the Taylor & Francis Group, an informa business

Publisher's Note
The publisher has gone to great lengths to ensure the quality of this reprint but
points out that some imperfections in the original copies may be apparent.

British Library Cataloguing-in-Publication Data
A catalogue record for this book is available from the British Library

Library of Congress Cataloging-in-Publication Data
A catalog record for this book has been requested

Typeset in Times New Roman
by Apex CoVantage, LLC

ISBN 13: 978-1-03-209321-5 (pbk)
ISBN 13: 978-1-138-08885-6 (hbk)

To my loving parents.

Contents

Tables

Abbreviations

ABS	Nagoya Protocol on Access to Genetic Resources and the Fair and Equitable Sharing of Benefits Arising from Their Utilization to the Convention on Biological Diversity
ARPANET	Advanced Research Projects Agency Network
CBD	Convention on Biological Diversity
CEC	Commission on Education and Communication (IUCN)
CEESP	Commission on Environmental, Economic, and Social Policy (IUCN)
CEL	World Commission on Environmental Law (IUCN)
CEM	Commission on Ecosystem Management (IUCN)
CI	Conservation International
CIEL	Center for International Environmental Law
CITES	Convention on International Trade in Endangered Species of Wild Fauna and Flora, also known as the Washington Convention
CMS	Convention on Migratory Species
COP	Conference of the Parties
CSD	United Nations Commission on Sustainable Development
CSO	Civil society organization
CSR	Corporate Social Responsibility
ECOSOC	United Nations Economic and Social Council
EDF	Environmental Defense Fund
EU	European Union
FAO	Food and Agriculture Organization
FIELD	Foundation for International Environmental Law and Development
FSC	Forest Stewardship Council
GA	General Assembly
GEMS	United Nations Global Environment Monitoring System
GESAMP	Group of Experts on the Scientific Aspects of Marine Environmental Protection
GMO	Genetically Modified Organism
HQ	Headquarters
HTPP	Hypertext Transfer Protocol
IAEA	International Atomic Energy Agency

ICANN	Internet Corporation for Assigned Names and Numbers
ICTs	Information and communication technologies
IGO	International governmental organization
ILC	International Law Commission
IMO	International Maritime Organization
INGO	International non-governmental organization
IO	International organization
IP	Internet Protocol
IPCC	Intergovernmental Panel on Climate Change
ISP	Internet Service Provider
IT	Information technology
ITU	International Telecommunication Union
IUCN	International Union for Conservation of Nature
MDG	Millennium Development Goals
M&E	Monitoring and Evaluation
NGO	Non-governmental organization
NRDC	Natural Resources Defense Council
OECD	Organization for Economic Co-Operation and Development
R&D	Research and Development
SCAR	Scientific Committee on Antarctic Research
SCLDF	Sierra Club Legal Defense Fund
SCOPE	Scientific Committee on Problems of the Environment
SCOR	Scientific Committee on Oceanic Research
SDG	Sustainable Development Goals
SSC	IUCN Species Survival Commission
TLD	Top-Level Domain
TNC	The Nature Conservancy
UN	United Nations
UN-REDD	United Nations Programme on Reducing Emissions from Deforestation and Forest Degradation
WCC	World Conservation Congress
WCED	World Commission on Environment and Development
WCPA	World Commission on Protected Areas
Web	World Wide Web
WEF	World Economic Forum
WHC	World Heritage Committee
WMO	World Meteorological Organization
WSSD	World Summit on Sustainable Development
WWF	World Wide Fund for Nature

Acknowledgments

Several people played an important role in the accomplishment of this book. First, I would like to thank Prof. Josep Ibañez of the Pompeu Fabra University for his continuing assistance and support over the last years. I would also like to thank the Routledge publishing team for their flexibility and great support, Prof. Dusan Sidjanski, and my friends and colleagues at the Centre de Compétences Dusan Sidjanski en Études Européennes, at the Global Studies Institute of the University of Geneva, and at the Graduate Institute of International and Development Studies. I am also deeply grateful to my partner, whose presence made this book possible. To my parents for their love throughout my studies and beyond.

Introduction

Nature sustains all aspects of human life and provides food, medicine, innovation, recreation, and shelter. Although often considered an inalterable, everlasting given, the earth is in a continuous process of adaptation to human needs, and suffers more and more losses in terms of species, ecosystems, and beauty. Demographic and economic booms of recent decades have put pressure on the earth's natural resources and capacity for adaptation. As indicated in the Synthesis Report of the Intergovernmental Panel on Climate Change (IPCC)[1] warming of the climate system is unequivocal, and many of the observed changes such as warmer atmosphere and oceans, diminished amounts of snow and ice, or higher sea levels happen at an alarmingly fast pace.

As climate change is one of the greatest menaces to people and organizations throughout the world, its impact on sustainable development and the achievement of the Sustainable Development Goals (SDGs) is also substantial. Since there is common agreement that on average global temperatures have risen by 1 degree Celsius compared to pre-industrial levels, the evolution of climate change, and whether global temperatures will rise more than 2 degrees Celsius in the near future, represents a factor of incertitude and will contribute to determining the success of the implementation of development initiatives, in particular for the poorest and most vulnerable. The three post-2015 agendas for action – the Paris Agreement, the 2030 Agenda for Sustainable Development, and the Sendai Framework for Disaster Risk Reduction – highlight the interdependence of all environmental and development issues and the subsequent need to adopt a holistic approach when it comes to global environmental governance.

At the closing ceremony of the 2018 Global Climate Action Summit in San Francisco, UN Climate Change Executive Secretary Patricia Espinosa recalled the past of the city that saw the birth of the United Nations system, and made specific reference to the American President Harry Truman, who told delegates when signing the UN Charter: "You, members of this Conference, are to be the architects of the better world. In your hands rests our future."[2] If new transboundary challenges such as environmental issues put pressure on the nation-state system,[3] states have a crucial role to play, not only to tackle global environmental issues and to protect global public goods, but also to encourage other stakeholders to take action. Since the creation of the United Nations system, states have developed a wide array of agreements to manage global public goods and in particular the environment. From the United Nations Framework Convention on Climate Change and more recently the Paris Agreement, to the Convention on Biological Diversity (CBD), global environmental governance has become the complex web of agreements and policies we know today.

The governance of the environment reflects quite well the diversity of views related to the environment. Contrary to government, which refers solely to the role of states, the concept of governance encompasses a much broader understanding of the international relations spectrum and "relates to any form of creating or maintaining political order and providing common goods for a given political community on whatever level."[4] Global environmental governance is indeed multi-actor and multilevel.[5] It counts multiple actors, including states, intergovernmental organizations, businesses, and a large and diverse civil society group that includes a myriad of non-governmental organizations, scientific expert networks, academia, local communities, indigenous people, and their communities, to cite but a few. Their respective roles have evolved over past decades with an increasing involvement of civil society on the international stage.

In her speech, Executive Secretary Espinosa further acknowledged the role of non-state actors in overcoming the environmental challenges ahead of us, and highlighted the need to address climate change through an inclusive global collaboration that she called inclusive multilateralism, which includes "the participation of subnational institutions such as states, provinces, cities and regions, as well as businesses, investors and non-governmental organizations."[6] Indeed, the solutions to these numerous environmental challenges are to be found not only in agreements and policies, but also in an inclusive governance format, which would allow the participation of those most affected by environmental degradation. By including a large array of actors, governance processes can benefit from the diversity of knowledge and expertise and from the social legitimacy provided by a large participation base. Hence, as John G. Ruggie argues, "governance, at whatever level of social organization it may take place, refers to conducting the public's business – to the constellation of authoritative rules, institutions, and practices by means of which any collectivity manages its affairs."[7]

Some global decision-making processes and institutions are sometimes perceived as being quite distant from the people who experience the implications of such agreements or their absence. Civil society is probably the most vocal conglomerate of actors to demand more open and participatory global governance mechanisms. The performance of such mechanisms is intrinsically linked to their local and national roots: the performance of a global measure to protect the environment depends on the level of participation and ownership of other actors, in particular at the local level. Therefore, the question of the social legitimacy of global environmental governance mechanisms is linked to their effectiveness and the success of their implementation. Inclusive participation to overcome global environmental challenges is at the heart of this book.

Thanks to the generalization of new information and communication technologies (ICTs), most actors on the international stage have gained new skills. Much has been written on how the internet, mobile phones, or social media, among other technologies and practices, have over the years become key tools for all actors in international relations. States develop social media communication campaigns to influence the nationals from other countries through public diplomacy initiatives. International organizations use mobile tools to communicate their activities. Scientists and academics collaborate and exchange ideas and knowledge through web-based scientific platforms. The implications of the generalization of current and emerging forms of ICTs are part of a vast change of paradigm where new technologies are only one factor of change among others. The generalization of new ICTs has led to new organizational principles in world politics.[8]

ICTs have given access to information to a wide spectrum of actors and individuals on the planet, thereby reducing the traditional informational asymmetry between governmental representatives and local activists for instance, or between organizations and individuals. The internet and social media have reduced communication and coordination costs dramatically, allowing non-state actors and in particular civil society to acquire new skills to inform, consult with, and mobilize their audiences, and by doing so, raise awareness about environmental issues, shame actors for their actions, defend specific policies, and organize protests to put pressure on governments and thus influence how the international agenda is set. Since these technologies are affordable and highly efficient in terms of outreach, by providing access to global audiences, they were rapidly embraced by individuals and local communities, leading to the emergence of new forms of activism and lobbying. For the same reasons, ICTs have become a favored instrument for civil society organizations (CSOs) to gain additional influence on the international stage by increasing their visibility, improving their coordination capacity, and refining their fundraising strategies.

In addition, ICTs not only provide new instruments and outreach channels to communicate and influence others, but also transform conservation techniques and outcomes. Emerging technologies such as blockchain, big data, and artificial intelligence (AI) are opening new horizons. Unforeseen until recently, big data and AI technologies allow the collection and analysis of vast amounts of data, in terms of both geographic and time scope; they also decrease the costs associated with project management and the need for human resources for basic tasks such as image recognition. Blockchain enables the development of more transparent, inclusive, and efficient decision making and governance procedures on a large scale. These are some of the numerous innovative applications stemming from these three emerging technologies that this book will develop in the following chapters.

Moreover, the generalization of ICTs has contributed to the emergence of new forms of authority on the international stage. Since civil society has gained new skills in terms of outreach and coordination, but also in terms of data collection and analysis, and efficient governance, its participation on the international stage is evolving. If their status often remains confined to the realm of observers in multilateral negotiations, civil society organizations are increasingly recognized as new forms of authority, in particular when it comes to scientific expertise, local knowledge, diversity representation, and project implementation, among others. Furthermore, ICTs allow the emergence of new models of coordination and governance mechanisms, allowing iconic movements such as 99% or Indignados to emerge and grow rapidly and internationally without the need to institutionalize.

The role of civil society on the international stage is becoming as significant today as the role of states was in the past century.[9] Indeed, in terms of functions, civil society contributed to global environmental governance through advocacy groups (agenda setting, influencing the outcome of international negotiations), knowledge brokers (field expertise, scientific knowledge), service providers (implementation), and watchdogs.[10] The adoption of ICTs by a large number of civil society organizations has provided them with new tools that foster their functions in global environmental governance. Although they are not the only actors to consider, they offer prominent illustrations of the use of ICTs in a global governance context, and more precisely how technology could help overcome some limitations such as restricted access to decision-making processes and limited financial and human resources. In that sense, this book discusses how ICTs can support the participation of civil society in global environmental governance and achieve Sustainable Development Goal 17 entitled "Revitalize the global partnership for sustainable development."[11]

While ICTs are empowering tools, they also present some concerns. First, the notion of digital divide that describes the unequal access to technology in the world still represents a challenge. Rates of internet penetration for instance are substantially different from one region to the next. Moreover, ICTs are increasingly (and rightly) perceived as a menace by states and other international actors, including civil society. Indeed, in this domain, offenses are always being committed, barriers to enter low, and deterrence extremely difficult to implement since attribution is complicated.[12] ICTs are used by criminal groups to conduct illegal operations and fraudulent activities, and by some states to censor content, control information, identify and isolate opponents, reduce access to specific information, and spy on parts of the population. They are also used to disseminate fake news and influence the outcome of democratic processes. ICTs have even become weapons of mass destruction with the example of viruses that can affect a nuclear electric power plant. However, although these questions are of significant concern, these aspects of ICTs are not the topic of this book.

This book focuses on the use of current and emerging ICTs by environmental civil society organizations. Its main objective is to understand how these digital technologies support the participation of CSOs in global environmental governance. Participation is understood in a broad sense, and encompasses all activities CSOs perform to take part in global environmental governance, including advocacy, making proposals, knowledge production and distribution, monitoring, field project implementation, fundraising, and promoting sustainable behaviors among citizens. All these activities benefit from the use of current and emerging digital technologies. An increasing number of CSOs are developing technological expertise that supports their participation in global environmental governance. The first four chapters focus on contemporary forms of ICTs, while the last three discuss emerging technologies.

The first chapter introduces some key elements that will be further developed in subsequent chapters, in particular the importance of digital technology and civil society in global environmental governance. It begins by discussing how these two elements are at the core of the concept of sustainable development, and it then examines the emergence of the global civil society and new forms of authority on the international stage. It further discusses the information society, and then examines how civil society organizations and the environmental community at large benefit from the use of ICTs. Without the adoption of efficient and affordable digital technologies that allow access to information, collaboration, and participation in policy- and rule-making processes, global environmental governance would probably be very different today and include fewer actors, in particular civil society actors.

The second chapter focuses on websites as the primary instrument used by CSOs to acquire visibility and credibility with the general public, provide information about their activities, and convey their messages to influence the international agenda. In other words, websites are instruments of advocacy, in the sense that they allow wide dissemination of information as well as interaction with the audience. In that context, websites offer unprecedented opportunities to reach two advocacy targets, the general public and state representatives, by providing broader access to information, consulting with their base audience, and mobilizing their stakeholders to act. This chapter first briefly presents the methodology used, and then discusses how a sample of CSOs accredited by the United Nations Environment Assembly (UNEA) use their websites to provide information, consult with their audiences, and mobilize them to support their advocacy strategies and promote their cause on the international stage.

The third chapter focuses on social media platforms, and how CSOs develop their presence on these platforms to advance their advocacy strategy. Hence, social media has profoundly transformed advocacy work by providing a new array of tools and communication channels to reach out to a wide global audience. CSOs accredited by UNEA use their social media platforms to set the agenda according to their interests and objectives. Their advocacy tactics depend largely on their mission. The choice of social media tactics, tone of voice, regularity of publishing, and story line depend on how they are positioned in global environmental governance. This chapter first presents some specific considerations about social media platforms, and then analyzes the advocacy strategies of a sample of CSOs accredited by UNEA on Facebook and Twitter.

While the previous two chapters focus on advocacy strategies, the fourth chapter examines how ICTs support the development of a multi-stakeholder governance process that allows CSOs to make proposals for the protection of the environment. It examines the multi-stakeholder decision-making process developed by the International Union for Conservation of Nature (IUCN), one of the oldest global environmental organizations. This – almost unique – governance mechanism, which provides the same voting rights to state and civil society Members, is at the heart of the organization and enables Members to make proposals in the field of nature conservation at large. For decades, non-governmental Members have been able to vote on and participate in core governance processes on an equal footing with governmental actors. This chapter discusses the progressive digitalization of the governance processes. The use of internet and web platforms allows Members to initiate proposals and vote on them, thereby greatly facilitating the process. This chapter first discusses the process in detail, and then analyzes the use of ICTs and their impact on the participation of various governmental and non-governmental stakeholders. The opportunity to evaluate how ICTs affect the participation of CSOs in a multi-stakeholder process is quite unique and provides an illustration of the potential of digital technologies when it comes to improving representative decision-making processes on the international stage.

The fifth chapter examines an emerging[13] technology, namely blockchain, and how it is used by CSOs to fundraise, promote sustainable behaviors among citizens, and monitor other actors of global environmental governance. This global digital ledger, which stores any type of transaction between two entities or individuals in a certifiable and enduring way, offers new and attractive solutions in terms of increased trust, financing sustainability, transparency, incentives, and distributed governance. Good governance is key to managing natural resources sustainably. Blockchain provides a new set of skills for CSOs to become more efficient and more effective. According to the UN Climate Change secretariat (UNFCCC), blockchain technology can contribute to enhanced climate action and sustainability, in particular in terms of trust, transparency, incentive, and financing.[14] This chapter first explains how blockchain technology works, and then presents some of its main applications by environmental civil society organizations.

The sixth chapter looks at another emerging technology in the field of global environmental governance: big data. We live in a world of data. Today more than ever, data are at the center of the economy. Individuals, organizations, and governments consume and produce large amounts of data. If the concept of big data aims to describe how large and complex data sets are analyzed[15], it also describes how organizations collect data from new sources. This chapter provides an overview of the current and potential opportunities that big data offer in the context of global environmental governance. It will first discuss the concept of big data and then discuss its applications in fields related to global environmental governance, focusing on knowledge production and distribution. Big data technologies cover extensive spatial scales and a wide range of scientific fields. Hence organizations can identify trends and patterns in the complexity of human behaviors and the diversity of species and ecosystems.[16] They can also develop prediction models based on facts and data.

The final chapter focuses on artificial intelligence (AI). AI is a general-purpose technology, similar to electricity in the sense that its applications are endless, and it will affect society at large. Artificial intelligence holds many promises in global environmental governance, and in particular for CSOs to enhance their knowledge production and field implementation capacity. AI can help in predicting the effects of climate change and the path of storms, monitoring the spread of diseases, understanding human decision-making processes in terms of energy consumption and protection of the environment, or identifying areas most in need of attention. AI is increasingly taking a leading role in making sense of the large volumes of data recorded by satellites, drones, smartphones, and sensors throughout the planet. This new technology already allows us to better understand our planet, from its urban communities to its most remote and untouched areas, from high mountain pikes to deep oceans. This chapter first examines the concepts of artificial intelligence, algorithms, and machine learning, then addresses the problems AI can solve, and finally discusses how AI supports the participation of CSOs in global environmental governance, and more precisely in their knowledge production and field implementation capacities.

By examining some of the most-utilized current digital technologies, and presenting some of the most prominent emerging ones, this book aims to illustrate how active civil society organizations operate, and how ICTs support some of their roles, and therefore their participation in global environmental governance.

Notes

1 Pachauri, R.K., Meyer, L.A., 2014. *Climate change 2014: Synthesis report. Contribution of working groups I, II and III to the fifth assessment report of the intergovernmental panel on climate change.* Geneva, Switzerland: IPCC, p. 2.
2 Espinosa, P., 2018. *We must increase climate action and create unstoppable momentum towards COP24.* UN Climate Speech. https://unfccc.int/news/we-must-increase-climate-action-and-create-unstoppable-momentum-towards-cop24
3 Held, D., 1995. *The anarchical society: A study of order in world politics.* London, UK: Palgrave Macmillan, p. 74.
4 Risse, T., 2004. Global governance and communicative action. *Government and Opposition*, 39(2), p. 288.
5 Andonova, L.N., Betsill, M.M., Bulkeley, H., 2009. Transnational climate governance. *Global Environmental Politics*, 9(2), p. 52.
6 Espinosa, P., 2018. *We must increase climate action and create unstoppable momentum towards COP24.* UN Climate Speech. https://unfccc.int/news/we-must-increase-climate-action-and-create-unstoppable-momentum-towards-cop24
7 Ruggie, J.G., 2004. Reconstituting the global public domain: Issues, actors and practices. *European Journal of International Relations*, 10(4), p. 504.
8 Castells, M., 2012. *Networks of outrage and hope: Social movements in the internet age.* Cambridge, UK: Polity Press, p. 227.
9 Edwards, M., 2014. *Civil society.* Cambridge, UK: Polity Press, p. 2.
10 Marchetti, R., 2010. *The role of civil society in global governance: Report on the joint seminar organised by the EUISS, the European Commission/DG research, and UNU-CRIS.* European Union Institute for Security Studies, p. 3. www.iss.europa.eu/sites/default/files/EUISSFiles/Civil-Society_Report.pdf
11 Xprize. *Op Cit.*
12 Deibert, Ron, 2012. *Big data meets big brother: The political economy of cyber security.* Video presentation at Watson Institute for International Studies, Brown University. http://watson.brown.edu/events/2012/ronald-deibert-big-data-meets-big-brother-political-economy-cyber-security [Accessed 20 July 2013].
13 Emerging technology is understood here as a technology that has recently become public knowledge, and is increasingly used by organizations in the field of global

environmental governance. Contrary to websites and social media platforms for instance, blockchain, big data, and artificial intelligence are not yet generalized (at various degrees) among environmental actors.

14 UNFCCC, 2018. *UN supports blockchain technology for climate action.* See more at https://cop23.unfccc.int/news/un-supports-blockchain-technology-for-climate-action
15 Oracle, 2018. What is big data? *Oracle.* www.oracle.com/big-data/guide/what-is-big-data.html
16 Death, R.G., 2015. An environmental crisis: Science has failed; let us send in the machines. *Wiley Interdisciplinary Reviews: Water*, 2(6), p. 596.

1 ICTs and the emergence of environmental civil society organizations

1. Introduction

In Europe, the first information revolution occurred in 1450 when Gutenberg invented the printing press. Beforehand, the Catholic Church controlled the production and diffusion of most written content. Books were to be found only in monasteries, and therefore only a very few intellectuals and clericals were literate. Decades later, hundreds of thousands of books from bestselling authors such as Luther and Erasmus were produced and sold in their lifetime.[1] The invention of the printing press triggered new needs in terms of rules and laws, such as censorship and copyright. Thanks to these new rules, newly born states gained power over the Catholic Church. Indeed, the way information is produced, managed, and distributed can lead to political, social, and organizational changes.[2]

This new technology of production and distribution of information led to the emergence of the Renaissance period, and the flourishing of knowledge, innovation, and literature. New scientific ideas questioned how the world was perceived and proposed new perspectives. This intellectual ebullition and the possibility of sharing information more easily than before contributed to the future economic, military, and intellectual predominance of Europe.

Similarly, a series of transformations have taken place over recent decades, at society and organization levels, thanks to the generalization of the internet, and more generally of information and communication technologies (ICTs). Among the many changes, the internet and ICTs[3] in general have undeniably empowered individuals and grassroot organizations with new communication, collaboration, and organizational competences. The internet for instance allows organizations to share information instantaneously with the entire world, to coordinate local to global actions affordably, and to include a wider audience of stakeholders in their decision-making processes, increasing their legitimacy naturally.

The internet is also increasingly used through smartphones that empower each individual with a high-capacity information and communication device connected to the world. This is the era on which this book primarily focuses, and it is the one that has seen the most transformations in global environmental governance with the emergence of new interconnected, well-informed, sometimes politicized and globalized public spheres. The increasing use of ICTs, and in particular social media, in civic interactions is probably the most debated civil society phenom-enon in recent years.[4]

This first chapter aims to introduce some key elements that will be further developed in the next chapters, and in particular the changing role of digital technology and civil society in global environmental governance. It will start by discussing how these two elements are at the core of the concept of sustainable development, which later led to the adoption of the Sustainable Development Goals (SDGs),and then will examine the emergence of the global civil society and new forms of authority on the international stage. It will further discuss the infor-mational society, and then examine how civil society organizations (CSOs) and the environmental community at large benefit from the use of ICTs. Without the adoption of increasingly efficient and affordable digital technologies that allow access to information, collaboration, and participation in policy- and rule-making processes, global environmental governance would probably be very different today and include fewer actors, in particular fewer civil society actors.

2. Emergence of new forms of authority

From studying nature to global environmental governance, decades and centuries have passed. Indeed, studying nature is not a recent activity. Already in ancient Greece, Aristotle referred to it.[5] Scientists such as Alexander von Humboldt explained the impact of the environment on plants,[6] Thomas Malthus explained the links between demography and natural resources,[7] and Charles Darwin pro-posed his theory of evolution.[8] In 1866, Ernst Haeckel used the word "ecology" for the first time to discuss the relation between one element of nature and its envi-ronment.[9] In 1864, in his book *Man and Nature*, George Perkins Marsh[10] argued for the first time that Man was the main cause of the extinction of species.

It is generally agreed that 1972 and the United Nations Conference on the Human Environment mark the beginning of the modern form of environmen-tal conservation that includes global governance mechanisms. According to the Commission on Global Governance:

> Governance is the sum of the many ways individuals and institutions, public and private, manage their common affairs. It is a continuing process through which conflicting or diverse interests may be accommodated and cooperative action may be taken.[11]

Today, global environmental governance is characterized by the participation of a large number of actors, including the private sector, intergovernmental and non-governmental organizations, groups of experts, and international courts, to cite only a few. Global environmental governance is also characterized by the capacity of this multitude of actors to influence policy- and rule-making processes in vary-ing degrees. The emergence of civil society organizations as a group of actors on the international stage is understood here as the global civil society.

2.1. Civil society, technology, and sustainable development

The concept of sustainable development was first coined in 1987 in the report entitled *Our Common Future*[12] by the World Commission on Environment and Development (WCED), also called the Brundtland Commission in reference to the former Norwegian Prime Minister Gro Harlem Brundtland, Chair of the Commission. This report was influenced by the 1980 World Conservation Strategy,[13] prepared by the International Union for Conservation of Nature (IUCN), with the collaboration of the United Nations Environment Programme (UNEP), the World Wildlife Fund (WWF), the Food and Agriculture Organization of the United Nations (FAO), the United Nations Educational, Scientific, and Cultural Organization (UNESCO), governments, civil society organizations, and the scientific community, all arguing for a sustainable approach to development that would support conservation rather than deter it.[14]

The objective of the Brundtland Commission was to formulate long-term environmental strategies to achieve a sustainable form of development in the following decades by envisaging development not only from an economic perspective but using a more holistic approach that would include social and environmental dimensions also. These three pillars constitute the well-known definition of sustainable development from the previously mentioned report: "Sustainable development is development that meets the needs of the present without compromising the ability of future generations to meet their own needs."[15] This report also called for the participation of non-state actors in achieving sustainable development, and in particular recognized the role of citizen groups, civil society organizations, academia, and the scientific community in raising awareness about development, social, and environmental issues. Moreover, the report acknowledged the need to include these stakeholders in decisions about resource allocations for the additional legitimacy and local expertise they can bring.

Another notion in the *Our Common Future* report beckons for our attention: the use of technology to achieve sustainable development objectives. The internet and other recent forms of information and communication technologies were either not invented or not yet widely used, but it is worth noting that on multiple occasions, the authors highlighted the importance of reorienting technology toward sustainability and sharing knowledge and technologies between developed and developing countries. Technology is understood here in a very broad sense, and includes information systems such as satellite observation, but also innovations in terms of food processing and distribution, energy consumption, pollution reduction, and agriculture, among others. This leads one to conclude that technology and knowledge exchange are at the core of the concept of sustainable development.

The report, which was welcomed by the UN Assembly in its Resolution 42/187,[16] led to the 1992 United Nations Conference on Environment and Development (UNCED), also called the Earth Summit or the Rio Conference. States agreed on a common Declaration[17] with 27 principles that provide a guideline for the implementation of a sustainable future. In this Declaration, technology and knowledge are also recalled as important factors to achieve sustainable development. For instance, Principle 9 highlights the importance of sharing scientific and technological knowledge, and the need to develop new and innovative technologies. Moreover, four Principles underline the role of civil society, women, youth, and indigenous people in shaping a sustainable future. More precisely, Principle 10 recommends that states include all concerned citizens in decision-making processes to handle environmental issues, and ensure public awareness and access to environmental information. Principles 20 and 21 acknowledge the implication of women and youth in environmental management, while Principle 22 urges states to recognize the identity, culture, and traditional knowledge of indigenous people, their communities, and other local communities, which are essential to achieving sustainable development.[18]

Agenda 21,[19] another outcome of the 1992 Earth Summit, also acknowledges a multi-stakeholder approach to achieving sustainable development objectives. In its Section III entitled "Strengthening the role of major groups," it recognizes the following partners: women, children and youth, indigenous people and their communities, non-governmental organizations, local authorities, workers, business and industry, the scientific and technological community, and farmers. Moreover, it recognizes the role of relevant and usable information for decision-making processes, and the importance of access to information for all: "The need for information arises at all levels, from that of senior decision makers at the national and international levels to the grass-roots and individual levels."[20] It also mentions the necessity to establish and strengthen "electronic networking capabilities."[21] Although the internet was used only in a few places, and social media platforms did not yet exist, it is important to highlight that the role of information and technologies was already identified in 1992 as being crucial for development. These items can be seen as the predecessors of future recommendations to use ICTs for development.

The 1992 Earth Summit also led to the adoption of the Non-Legally Binding Authoritative Statement of Principles for a Global Consensus on the Management, Conservation, and Sustainable Development of All Types of Forests.[22] The Forest Principles for short recommend in Principle 2(d) that states promote the participation of all stakeholders when developing forest policies, including local communities and indigenous people, industries, labor, non-governmental organizations and individuals, forest dwellers, and women.[23]

Lastly, at the 1992 Earth Summit, three legally binding agreements, also referred to as the Rio Convention, were opened for signature: the United Nations Framework Convention on Climate Change (UNFCCC), the Convention on Biological Diversity (CBD), and the United Nations Convention to Combat Desertification (UNCCD), all of which aim to foster the participation of interested parties, including civil society organizations as mentioned in article 4(i) of the UNFCCC,[24] the Preamble of the CBD,[25] and the Preamble and articles 3(c) and 5(d) of the UNCCD.[26]

At the 1992 Earth Summit, the United Nations Commission on Sustainable Development (CSD) was established to coordinate, through yearly meetings, the work of the organization following the conference. In 1993, civil society organizations and other major groups convened to coordinate their actions. The United Nations Non-Governmental Liaison Service (UN-NGLS) organized a series of follow-up regional phone conferences and a meeting for those attending the 1993 United Nations Climate Change Conference in Copenhagen. A year later, it established the NGO Steering Committee to the United Nations Commission on Sustainable Development (CSD).[27]

Another milestone in global environmental governance is the 2000 Millennium Summit, which led to the adoption of the Millennium Declaration[28] from which stem the Millennium Development Goals (MDGs).[29] Goal 8 entitled "Develop a Global Partnership for Development"[30] acknowledges the importance of information and communication technologies, and in particular mobile communication (2G) and internet penetration in Target 8.F. The year 2000 corresponds to a milestone for ICTs also; not only is it considered to be the year when the internet started to become generalized, but the beginning of the 21st century also heralded the birth of social media. It is not surprising therefore to see a reference to these technologies here. In terms of civil society involvement, the Millennium Summit followed a two-year consultation period with the participation of more than 1000 civil society organizations from more than 100 countries, which published in May 2000 "We the Peoples Millennium Forum Declaration and Agenda for Action."[31] In this Declaration, civil society organizations reaffirm their role in achieving sustainable development objectives, and propose a set of actions for the United Nations, governments, and CSOs to eradicate poverty, ensure peace, security, and disarmament, reduce the negative impacts of globalization, defend human rights, support sustainable development and the protection of the environment, and strengthen and democratize the UN and IOs.[32]

In 2002, the World Summit on Sustainable Development (WSSD) in Johannesburg once more recognized the role of civil society and other major groups in achieving sustainable development objectives. The WSSD Plan of Implementation[33] reiterates the intention to develop partnerships between governments and non-governmental organizations[34], and the necessity to support youth and youth councils.[35]

At the 2012 United Nations Conference on Sustainable Development, also called the Rio+20 Conference, the outcome document entitled "The Future We Want" reaffirmed the importance of involving the major groups to achieve sustainable development objectives. States invited civil society organizations, local communities, volunteer groups and foundations, migrants and families, older persons, and persons with disabilities to take part in the decision-making processes related to sustainable development through consultation and collaboration mechanisms.[36] More precisely, the role of civil society was reiterated multiple times, but one particular aspect needs to be noted: the link between civil society engagement and information technologies.

We acknowledge the role of civil society and the importance of enabling all members of civil society to be actively engaged in sustainable development.

> We recognize that improved participation of civil society depends upon, inter alia, strengthening access to information and building civil society capacity and an enabling environment. We recognize that information and communications technology is facilitating the flow of information between governments and the public. In this regard, it is essential to work towards improved access to information and communications technology, especially broadband networks and services, and bridge the digital divide, recognizing the contribution of international cooperation in this regard.[37]

Indeed, local communities and civil society need to have access to information in order to take part in online participation mechanisms. It is an essential role of digital technologies to empower traditionally marginalized populations and provide access to data and information, which were previously limited to states and large organizations. Access to information is therefore an important aspect of sustainable development, and as mentioned previously, it needs to be associated with citizen engagement and multi-stakeholder governance processes. In this sense, digital technologies allow capacity building and enhance the role of civil society organizations and citizens.

In 2012, the UN Commission on Sustainable Development became the High-Level Political Forum on Sustainable Development (HLPF), where civil society and major groups received participatory opportunities. Since then, they can attend all official meetings of the HLPF, access official information and documents, intervene in official meetings, submit documents, make recommendations, contribute orally or on paper, and organize side events and round tables in cooperation with Member states.[38]

Three years later and after consultation with civil society and major groups, the 2030 Agenda for Sustainable Development was adopted at the 2015 United Nations Sustainable Development Summit. This global agenda for sustainable development builds on the Millennium Development Goals and consists of 17 Sustainable Development Goals and 169 targets, ranging from eradicating poverty to gender equality. This ambitious agenda and its implementation bring together multiple stakeholders, including civil society and major groups. In particular, Goal 17 and Targets 17.16[39] and 17.17[40] focus on multi-stakeholder collaboration, with public-private and civil society partnerships. Moreover, Targets 17.6,[41] 17.7,[42] and 17.8[43] highlight the need to provide access to science, technology, and innovation and to enhance knowledge sharing in the world, and they support the use of empowering technologies such as ICTs. Here as well, the link between sustainable development, civil society participation, and digital technologies is highlighted. ICTs are recognized as enhancing instruments that can support development and the participation of citizens and civil society in shaping and implementing sustainable development.

Further to the adoption of the 2030 Agenda for Sustainable Development, the HLPF meets every year and includes non-state actors in the follow-up and review processes. Collaboration between the United Nations and civil society and major groups is realized through the UNDESA/Division for Sustainable Development (DSD).[44]

The concept of sustainable development, arising in the 1980s, is at the center of contemporary global environmental governance. As illustrated earlier, technologies and civil society engagement have been at the core of this concept since its inception, and have influenced a wide spectrum of global to local environmental policies and agreements.

2.2. Global civil society

The international stage is permanently evolving thanks to a series of factors, one being the generalization of new technological capacities to acquire and retain power. The 1990s is the decade that saw an unprecedented increase in the number of CSOs in the world. This increase is due to several factors, including the generalization of information and communication technologies, a wider understanding of global interdependence and the need to protect global public goods, and the spread of democracies at that time.[45] New means of transportation, communication, and information production and distribution accompanied globalization and the rapid growth of the nonprofit sector. These three processes took place in parallel and fed one another. Without new technology, globalization and the emergence of the global civil society would have not been possible, or at least would not have occurred as quickly.

Global civil society is a heterogenous group of actors that emerged in reaction to the negative impacts of globalization on society and populations. ICTs helped the politicization of global issues and institutions, which triggered the emergence of the global civil society. In many cases, social movements emerge and grow rapidly, on an ad hoc basis, and in reaction to a specific event, decision, or actor, although opposition to extreme forms of capitalism and globalization and support for universal human rights are common to most of these movements. As Kaldor argues, civil society is "the medium through which one or many social contracts between individuals, both women and men, and the political and economic centers of power are negotiated and reproduced."[46] It is also the product of nation-states and capitalism, in the sense that it arose spontaneously to mediate conflicts between the market economy and social life. In addition, it is a universal collective expression of individuals; it can be found in all countries and stages of development, although with various cultural expressions.[47]

Global civil society is a concept widely discussed among international relations scholars. It encompasses the organized social life or civil activity of individuals and collective actors in pursuit of various global to local political and nonpolitical goals. The global civil society inhabits the space between the private sector economy and the state. It includes a wide variety of actors with sometimes conflicting objectives: formal representative organizations such as parties, churches, lobbies, or trade unions cohabit with informal functional organizations such as charities, universities, think tanks, mass media; and with more informal social and political entities and their networks such as social forums, ad hoc activist coalitions, diasporas, networks, causes, or internationally coordinated social movements.[48] Civil society organizations in global environmental governance are equally diverse and represent a large range of perspectives and interests: "Global civil society is diverse, creative and chaotic. That's what makes it always interesting, often unpredictable, and sometimes very powerful."[49] Indeed they play a key role on the international scene; by raising awareness about a specific issue among the general public, CSOs strive to influence governments' actions and decisions.

Two important elements differentiate the global civil society conglomerate of actors from others: its voluntary nature (nonprofit organizations as opposed to multinational corporations) and civility (as opposed to terrorist groups that resort to violence for accomplishing their goals).[50] In addition, the "global" element of this concept refers to various types of entities: truly global associations such as the World Wide Fund for Nature (WWF) with audiences all over the world,[51] organizations with activities or audiences in various countries but that are not really global,[52] and local entities targeting global institutions in response to global issues.[53]

Furthermore, global civil society encompasses not only a large range of interests and perspectives, but also a wide array of governance structures, from membership organizations to foundations, and from information networks to "informal quasi-organizations."[54] It offers new models of governance that make great use of ICTs to cooperate and organize their work. In addition, social media platforms are used to reach out to their audiences at global, national, and local levels. The advantage is that the audience on social media and the organization with ICTs are both fluid and can evolve rapidly and at limited cost. They can adapt to a variation in funding, a new issue raised, and the involvement in a network of other organizations.

It is indeed undeniable that the emergence of the global civil society as an influential group of actors is linked to a dynamic process and the generalization of new ICTs, which drastically reduced the distance between people.[55] The advances in communication technologies have helped new organizations to emerge and others to grow rapidly beyond state boundaries.[56] Given their low cost, ICTs were rapidly adopted by civil society for multiple purposes: to raise funds and acquire new members, to organize obstructive actions and international events such as sit-ins or blockades, to simplify global coordination or to make it possible. Websites are also important for diaspora groups and INGOs.[57] Scholars focusing on social movements agree on the importance of these new forms of communication. Over the last decades, CSOs also became more and more professionalized, and included professionals and experts who worked on various environmental issues, coinciding not only with a growing awareness among populations of the importance of nature conservation, but also with a growing number of research projects and studies in this field.

Furthermore, the growing interconnectedness resulting from the generalization of the ICTs and the globalization process of trade and finance enabled the emergence of islands of civic engagement in countries suffering from military dictatorship. Civil society groups bypassed their national states to appeal to international networks and institutions. The relationship between local civil society groups and transnational networks and institutions contributed to the construction of a framework for global governance.[58]

Thanks to ICTs, civil society organizations have increasingly become global, challenging the global policies of international institutions and leading to the development of new accountability norms. They can offer more social legitimacy to global environmental governance mechanisms by representing local communities and being the voice of non-state and non-corporate actors. Many global institutions tend to collaborate with CSOs and civil society at large to gain additional social legitimacy and higher levels of public trust.[59]

Many studies confirm the influence of civil society organizations on global politics, and the emergence of CSOs as contesting authorities on the international stage, especially when setting agendas and monitoring results of policy implementation.[60] This is particularly true for their participation in the conferences of the United Nations.[61] Although often restricted to an observer status, CSOs use ICTs to raise awareness about specific issues and mobilize large-scale communities. In the digital age, when information and knowledge are the core of societies, the capacity to inform, consult, and mobilize represents power.

Another way in which the global civil society has an impact on states' decisions is with the production of specialized information about global issues such as the environment.[62] In addition to scientific expertise, CSOs have gained extensive field experience that they share with states and intergovernmental organizations. Part of this experience is accumulated through collaborating with a wide array of actors in the field.[63] In many regions of the world where the state is inefficient, CSOs are leading programs that provide basic services to populations.[64] They seem to be more flexible than government bureaucracies and therefore sometimes more efficient in delivering these services.[65] Some studies show successful examples of the authority of global civil society, especially in terms of monitoring human rights violations of states or generating international pressure.[66] States and IOs have become more conscious of the social and environmental aspects of their policies, programs, and decisions due to the work of civil society.[67]

A successful example of the existence, action, and influence of the global civil society is the "Save the Narmada Movement" in India. A dam project, co-funded by the World Bank in 1989, aimed to build 30 large, 135 medium, and 3000 small dams in the Narmada Valley in Madhya Pradesh and Maharashtra, flooding an estimated 120,000,000 hectares of land and displacing more than 300,000 people. In response to that threat, a group of indigenous people formed the Narmada Bachao Andolan (Save the Narmada Movement), which established links with many other Indian groups and international organizations.[68] Sixty thousand indigenous, landless laborers and peasants gathered to protest against the construction of these dams. Local protests were coordinated with political pressure by international non-governmental organizations such as the Environmental Defense Fund in Washington, Survival International in London, and Friends of the Earth in Tokyo. According to the World Bank official in charge of the project, the international networking was pivotal to the World Bank's cancelation of the project.[69]

The environmental movement became a global counterculture.[70] Meetings of the G8, the World Trade Organization, and the International Monetary Fund were targeted by grassroots protesters and green organizations in 1994 in Madrid, 1999 in Seattle, and 2001 in Genoa. The creation of the World Social Forum (the counterpart of the Davos Economic Forum) unified all forces acting to protect the environment, promote human rights, and fight against the negative impacts of capitalism.

The global civil society undeniably became a major group of actors on the international stage thanks to its new capacity to communicate and set the agenda. Its influence led to the adoption of new treaties, norms, and conventions, and to the creation of new organizations and programs to protect the environment. ICTs played a key role in enabling CSOs to gain this increased influence. ICTs also enabled the emergence of new forms of authority, and granted them additional visibility. In this new setting, states collaborate with new forms of authority to produce outcomes in nature conservation.

2.3. New forms of authority

Although states and IOs have recognized the benefits of a multi-stakeholder approach to managing global public goods since the early 1990s, CSOs have remained in many cases confined to the role of observers. For instance, Agenda 21 urged states to take measures to enhance the participation of civil society in international decision-making processes.[71] Principle 10 of the 1992 Rio Declaration states that environmental issues are best handled with the participation of all concerned citizens. This principle led to the adoption of the 1998 Aarhus Convention requiring parties to ensure access to information, justice, and public participation in environmental matters.[72]

CSOs have gradually developed new capacities and have become a key group of actors on the international stage: they implement field projects and provide services, produce scientific knowledge, monitor governmental actors and multinationals, advocate for new policies, take part in international negotiations, and raise awareness about issues related to the protection of the environment. These new roles were made possible thanks to the wide adoption of affordable means of communication, namely ICTs, that allow CSOs to coordinate their actions, share data and information, and communicate with their audience on a large scale. ICTs are not the only factor to explain the emergence of these new roles, but acted as an enabler. These new roles can also be understood as new forms of authority on the international stage, since CSOs are recognized as such by the other international actors. Indeed, states and intergovernmental organizations make regular use of their scientific reports, local expertise, monitoring, and project management competences to reach their development goals and advance some of their objectives on the international stage.

Authority consists of two elements: the capacity to induce deference on others, and the claim to be justified in doing so. What defines an authority is its capacity to make rules that are binding, perceived as such, and applied. Authority entitles an entity to make authoritative decisions in order to achieve an objective and to use coercive and persuasive powers. It contains two levels: level one prescribes rules, while level two enforces them.

Authority can stem from various sources: institution, delegation, principle, and expertise or capacity.[73] The institutional authority (or institution-based authority) comes from the institution an individual represents: the managing director of United Nations Environment (UNEP), the head of a governmental agency, or the Director General of an international NGO. This authority is limited to the functions of the institution it represents. The head of UNEP cannot pronounce policy on conflict resolution. The organization would lose its legitimacy.[74]

Delegated authority is authority that derives from another institution: states delegate some of their authority to implement development programs to IOs or NGOs for instance. NGOs increasingly play a significant role in the promotion and reporting of some universal values such as environmental conservation or human rights. They also implement some public tasks such as water distribution, infrastructure development, health protection, or education. This reflects the fact that some governments with limited means delegate part of their authority to NGOs. Indeed, NGOs are sometimes asked to play public roles to compensate for (global, regional, national, and local) governance gaps and failures.[75]

Principle authority is legitimate as long as it acts according to the principle. It loses its legitimacy otherwise.[76] Principle authority is based on widely accepted principles, morals, or values. It can be NGOs acting to save the planet, protect human rights, or defend victims of fraud. They call on altruism and act for the general good. For instance, a global network of 600 civil society organizations in 70 countries spread the word on the World Wide Web to defeat the Multilateral Agreement on Investment (MAI) in 1998. This is one of the most iconic examples of global civil society influencing global rule-making mechanisms to the point of preventing an agreement from happening. Its virtual presence and opposition, then supported by the world press, led to the withdrawal of this agreement from the international agenda.[77]

Expert or capacity-based authority stems from specialized knowledge: unlike the previous types, it belongs to the actor, for it is knowledge that grants the authority to make a judgment or take action. It is the capacity to perform a specific action (solving a common problem or providing a global public good) that justifies the capacity-based authority.[78] This form of authority is based on special knowledge or moral expertise. What is decisive for the epistemic authority is not the quality of the judgment or the rule produced but rather the quality of the reputation.[79] The source of this authority is what justifies the recognition of authority. The expertise is recognized by an audience as well-founded and impartial. For example, the IPCC is recognized as an authority on climate change thanks to the reputation it acquired while producing regular Assessment Reports.[80]

Increasingly, civil society organizations have produced some public goods. In Ghana, Zimbabwe, and Kenya, the nonprofit sector provides at least 40% of all healthcare and education services delivered.[81] Indeed, states and intergovernmental organizations often outsource part of their development work to civil society organizations. Due to their expert authority, stemming from their strong field experience and scientific knowledge, they are sometimes in a better position to implement international aid at the local level. Their field experience and scientific knowledge have become a precious source of information for states and IOs.

Often this source of authority has no means of enforcement, and, needing to avoid conflict, it must find other ways to persuade others to do what it wants. It therefore relies on the audience to appropriate the ideas emanating from the authority because of its recognized expertise. This is the case for the status of biodiversity and the Intergovernmental Science-Policy Platform on Biodiversity and Ecosystem Services (IPBES) for instance.

Expert authority is quite prominent in global environmental governance, for it embeds all organizations that produce scientific knowledge or are recognized for holding expertise in a certain field. The increasing development of indicators in global governance is an indication of this trend. This includes advice, recommendations, and the formulation of best practices. The production of knowledge and information is often seen as a central tool for inducing change in the behavior of institutions and individuals in a complex environment.[82]

These multiple sources of authority that are possible thanks to the adoption of ICTs have led to the development of a new concept of authority that encapsulates the evolution in global environmental governance: solid and liquid authority. Solid authority is all about command: the source of authority makes a rule and issues directives to its subjects, much like a monarch. In international relations, this represents only a small part of the authority in global governance today.

On the other hand, liquid authority accounts for the largest part of global governance, and in particular of global environmental governance. It recognizes the multitude of other practices of rule, which are informal (not formal or binding), ideational (not demanding), fleeting (not settled and fixed), and multiple (not unitary).[83] Liquid authority operates through soft law, recommendations, best practices, or policy dialogues. It takes into consideration the multiple forms of international private authority that are active on the international stage. It also describes the growing role of the global civil society in global environmental governance and the emergence of multiple authorities that compete for visibility, influence, decision-making power, and funds in global environmental governance.[84]

Another element of liquid authority is multiplicity. Whereas solid authority is unitary and centralized, liquid authority counts multiple sources of authority. In the case of climate change, multiple recognized sources of authority produce recommendations: governmental agencies, IPCC, UNFCCC, UNEP, WMO, IUCN, WWF, and other NGOs, as well as regional organizations such as the European Union, to cite only a few. This plurality of authorities means not only that authorities become "relative" in terms of level of influence, but that they also become relative to others and in interplay with others.[85]

Furthermore, liquid authority is not fixed: it is a constant process of change. One institution might be the main authority for some time and then disappear or be replaced by another. Authority can be shared among a wide range of institutions, and it can rest in this range of organizations rather than in one in particular.[86] This fleeting character implies that an institution can enjoy authority and be a recognized source of authority when created for a specific purpose. Then, when the need has disappeared, it can be dismantled.[87]

As discussed, ICTs allowed civil society organizations to become a crucial group of actors that can collaborate and compete with other traditional sources of authority. Civil society is now perceived as a crucial counterweight to state and corporate power, as it promotes and sometimes enforces transparency, accountability, and good governance.[88] For James Rosenau, world politics evolved and split into two: inter-state relations on one side, and on the other various non-governmental actors that are independent of the state-centric world but that often interact with counterparts in the state-centric world.[89]

Moreover, if CSOs have gained influence in the international arena, it is due in part to their growing number, but also to their capacity to become new forms of authority on the international stage. Thanks to ICTs, individual citizens can self-organize outside political circles and influence the conditions in which they live.[90] In that sense, civil society is the arena of contestation, debate, dialogue, and pluralism at local but also national and international levels (at different degrees). It is a source of both civility and incivility.[91] The emergence of the environmental movement in the 1960s, understood here as the group of civil society organizations supporting a cause related to the environment, makes no exception: it has become over the years a source of multiple forms of authorities. If ICTs did not trigger its emergence, they supported its global growth and transformation into a key actor on the international stage.

ICTs did not only have an impact of the emergence of multiple forms of authority and CSOs on the international stage, but also on society and individuals. Since digital technologies are present in people's everyday life, they tend to influence environmental actors – among which CSOs, but also the audience of these actors, in other words people and society. Since one of the main roles of CSOs is to raise awareness, it is crucial to understand how the recipients of these advocacy strategies have evolved, thanks to their adoption of ICTs.

3. Informational society

As mentioned previously, information revolutions are based on the emergence of new information technologies. The most recent one was triggered by the converging set of emerging technologies in microelectronics, computing telecommunications, and optoelectronics.[92] This new generation of technologies led to the emergence of a new paradigm, from which a new society that Manuel Castells names the "informational society" was born.[93] Although the digital divide describes the reality of many parts of the world without access to the internet, and remains an issue both internationally and domestically, internet penetration rates increase every year to a point where internet access is even sometimes considered a human right by some scholars and policy makers. The year 2000 is often used as a milestone to describe the emergence of web 2.0 and the era of social media. This is the era this book focuses on.

3.1. New patterns of interaction

Similar to the Renaissance thinkers who planted the seed for the European technological dominance that took place a few centuries later, the generalization of new ICTs in the 20th century contributed to the emergence of a new type of society in the 21st century.

This new society is based on a new paradigm, which consists of various elements. The first element is how technology acts on and transforms information. It is no longer solely new information that acts on technology as with previous industrial revolutions. It is also how information is becoming a raw material that can be used and reused for multiple purposes and in multiple aspects of life. Indeed, for the first time in history, the human mind is a productive force, and not just the decisive element of the production process.[94] Information itself becomes a resource to produce knowledge and wealth.

Second, ICTs are pervasive technologies. Indeed, information is an integral part of human life, and most processes of our individual and collective existence are influenced by new ICTs.[95] Third, ICTs favor networking in the shape of dynamic networks such as social media. They allow complex, global, and extensive interaction among individuals, which increases the potential for creativity and leads to unpredictable patterns of development. This leads to flexible processes and organizations that can change shape and form rapidly. Finally, the last element is the convergence of all technologies and media: microelectronics, telecommunications, optoelectronics, and computers are all integrated.[96]

This new paradigm has also an impact on how individuals and organizations interact: three new patterns of interaction have gradually emerged in the informational society: transparency, collaboration, and participation.

Transparency has indeed become a new pattern of interaction between individuals, but also between organizations.[97] The internet, social media, and mobile phones allow users to discuss among themselves worldwide and in real time with little constraint and few rules. Most discussions take place transparently in public and can be contributed to by every individual. This generation of ICTs also enables the general public to monitor public figures in an unprecedented way.[98] Until recently, the state controlled power with the help of surveillance techniques. This control is now reciprocal. The state seems more and more transparent to the public eye, as do the actions of political figures. A feeling of constant scrutiny might discourage illegal activities. The speed at which information circulates via the internet and mobile technologies is an additional deciding factor for the political elite when taking action.

Although more transparent political processes and decisions are positive in many instances, transparency is not ideal for all situations and entities. For instance, diplomatic relations need to keep a level of secrecy. Companies must ensure they keep their innovation advances safe from the competition. Individuals should be able to secure their personal information even when browsing the web. Most recent research shows that privacy breaches are what concern internet users the most.[99]

A second pattern of interaction is collaboration.[100] There is a new model of production in which people from all around the world join forces to produce information, create new online services, and innovate globally. ICTs have considerably accelerated the development of this second pattern of interaction thanks to voice over internet calls, instant messaging, video conferences, and cloud-based file sharing information systems for instance. Thanks to ICTs, it is nowadays easier and cheaper for an organization to coordinate individuals and initiatives throughout the world. Therefore states and multinational corporations are not the only ones to have this capacity but CSOs and individuals as well.

The final pattern of interaction is participation.[101] With the emergence of social media platforms, individuals are not only consumers of information and technologies. They have also the possibility to like, share, and comment on news items produced either by other individuals or by organizations (large and small, public and private). This new pattern of participation is linked to the two previous ones in the sense that citizens expect organizations and public institutions to provide a channel for them to express their views, which increases the level of transparency and the possibility of collaboration between individuals and organizations. In this context, ICTs reduce the gap between traditional forms of authority and individuals, increasing the institutional social legitimacy.

These new patterns of interaction lead to the emergence of new online public spaces, where individuals and organizations meet. As defined by Habermas, the public sphere is where common concerns are debated among citizens in an equal, free, and peaceful manner.[102] The capacity to deliberate democratically, to perceive shared interests, to cede territory to others, and to work together to achieve a common objective is key.

These new global spaces emerge on the international stage not only as a complex group of multilevel heterogeneous interests, but also as a platform: the technology itself that allows information to propagate faster, farther, deeper, and broader, as mentioned previously, is an integral component of this new public space. This new space of autonomy, which competes with other sources of authority to set the international agenda or push politicians and business leaders in one direction or another, is made up of people, individual social media users, but also the technology that was regulated by states and designed to foster this interconnection and free movement of ideas.

Although ICTs and in particular social media platforms are not exempt from critics as discussed in Chapter 3, they allow the emergence of new online spaces specifically for civil society to debate and protest. For instance, "Indignados" and "99%" are some iconic examples of such non-legislative and extra-judicial public arenas where participants debate societal and social issues, public policy, government initiatives, public goods, and cultural identity.[103]

Since the environment is a global public good, these new online spaces are the ideal places to debate its production and protection, as it concerns everyone from local to global. They are intensively used by CSOs to raise awareness and mobilize their audience as shown in the next chapters. In particular, social media platforms allow the emergence of online networks and communities that support the work of local and global CSOs and have developed the capacity to influence at various degrees some aspects of the international agenda. Information debated in these online spaces is heterogeneous with an infinite number of groups focusing on specific local to global causes and interests. This coexistence of an infinite number of online spaces represents a new feature on the international stage and was also triggered by the emergence of a new generation of users, born with these ICTs: the Net generation.

3.2. Net generation

The new patterns of interaction are largely embraced by the Net generation: children, teenagers, and young adults born since the generalization of new ICTs has integrated all aspects of the previously described new paradigm. And for the first time in history, these young generations know more about the dominant technology than their parents.[104] Indeed, they are used to looking up the information they need on the web: they are not just passively consuming, but active in customizing the knowledge they wish to receive through setting up their Twitter account, news feeds, blogs, or Facebook profiles.

Born with the third screen (the screen of mobile devices that came after television and computer screens), the Net generation is used to seeing electronically altered images or living in a virtual reality as experienced in online video games. Uniform products or services are no longer appropriate for these users, since they are used to tailor-made, real-time solutions that respond directly to their needs and desires. For example, they can choose to stream a TV series online instead of waiting for a channel to showcase it; they can customize or design their own clothes and order items from around the world; they can create communities of peers with the same interests and receive answers to their questions almost in real time.

This has led to a perceived need to be part of an online community to share feelings, values, and ideas. Institutions have become aware of this need and used them to support their interests. Obama for instance created an online platform for his two presidential campaigns where potential voters could discuss issues relevant to them and make their views known to the President. This openness and transparency were probably factors that led to his popularity among younger voters.

This Net generation has been provided with new ICTs, which broaden their perspective and motivate them to be politically active. The causes of the revolution are not new and stem from injustice, the lack of jobs, repression, violence, the non-respecting of human rights, or economic disparity. The tools used to fight this revolution are however no longer weapons, but the new ICTs, which allow the Net generation to raise awareness, denounce injustice, or call for help.

The internet and mobile phones are almost like the military tools used by the youth in some countries to spread the word, decide on a meeting point, join forces on an issue, locate snipers and send their location to friendly militant forces, and effectively change the society they are in. These newly created networks give a real sense of participation to the population. The clash is evident with authoritarian regimes and traditional forms of authorities such as governments, political parties, and religious figures. In this context, new ICTs created windows of opportunity for young people to take part in the destiny of their country; by being computer and internet savvy, the Net generation discovered its own way to be involved in politics and build a new society.

Over the past few years, the Net generation has been entering the workforce and joining organizations. This generation is larger than the baby boomers: 80 million compared to 78 million in the USA alone.[105] In other parts of the world such as Asia or Africa, the Net generation represents an even larger part of society. Demography is indeed another element to take into consideration when discussing the use of ICTs: North America and Europe count for less than 35% of all internet users; Asia counts for 45%, with a lower rate of penetration. This means there are vast parts of societies, in particular the youth in Asia and Africa, still to be connected. Although the internet may have emerged in western countries, its future might be decided elsewhere.[106]

The Net generation with ownership of online tools is becoming a powerful force to change societies. Thanks to education and an economy gradually organized around knowledge and information, information technologies have empowered the human mind[107] and dramatically expanded the horizon for creativity and innovation. ICTs have also provided new skills for the Net generation and others to organize protests and make their voice heard on the international stage.

3.3. Activism 2.0

Activism 2.0 describes the use of social media channels by activists for two main purposes: first to advocate (inform, consult, mobilize) and second to structure their movement differently, allowing the emergence of new governance models.

First, these new forms of activism create a new global public domain: an arena of debate, contestation, and action that revolves around the production and distribution of public goods.[108] Ideas spread faster when individuals are socially linked. Twitter, Facebook, and other social media platforms have increasingly become the favored space where citizens begin their participation in protests by either posting a photo, providing information for the press, or offering support for a cause by liking its page or following its hashtag.

The new forms of protests are cosmopolitan, modular, and autonomous. They are cosmopolitan, for people become aware of a wider community and not only the people they know. They are modular, for people can learn from others and understand their demands through new forms of communication. Moreover, they are autonomous since any individual can sign a petition or write a message on a blog, a forum, or a Facebook page.[109]

ICTs enable people to propose new content to the general public and invite anyone in the world who is interested into the heart of the conflict without any other interference or intermediary. Internet and mobile-based communication campaigns enable causes to become more visible and social ties between potential participants to be created away from traditional and sometimes censored media channels. Contrary to the situation before, social media and web 2.0 enable the production of new content directly by individuals.

Thanks to these new technologies, civil activists can easily take a snapshot with a mobile phone camera, record a short video, write a blog, send a tweet, or share comments on Facebook or Twitter. They can record a human right violation on their mobile phones and send it to their friends to post on the internet. This video is then seen by millions of people on both the internet and traditional channels such as TV. This is a good example of how new communication media have transformed activism: they transformed citizens into activists, who have access to global audiences and acquired much visibility. These media also give people more power, for they now have new surveillance tools that can be used against abusive political leaders.

These new forms of activism allow the Net generation not only to protest and communicate, but also to develop new models of governance according to the new patterns of interaction mentioned previously. Indeed, thanks to ICTs, alternative governance models have emerged without the necessity to become institutionalized.

Traditionally, when civil society movements grow, they become institutionalized. They take the familiar path from charisma to regularized routine, from inventiveness and passion to bureaucracy and hierarchy. Moreover, when they become more international they also become more institutional: in other words, an increased institutionalization supports the expansion of the movement. Institutionalization usually encompasses two aspects: internal institutionalization with professionalized management and a hierarchical structure, and external institutionalization with integration into institutional international processes or as a service provider for its members. In other words, the movement becomes an organization that either "does or provides something" on the international scene in a professional way. Over time, protesters are pacified and adapt when their requests are heard, and once they take part in a dialogue with authorities.

Thanks to ICTs, civil society movements can emerge online and remain global grassroot movements without the need to become institutionalized. Indeed, given that many social movements were founded by and for alter-globalization activists, their evolution did not follow the traditional pattern.

These new civil society movements have some common characteristics: they are organized and function as a network, and they make intense use of new ICTs such as internet-based and mobile communications. Although networks are old forms of social organization, they are now empowered by new ICTs, so that they can simultaneously combine flexible decentralization with focused decision-making processes.[110] Furthermore, mobile communications (call, SMS, or web) are not tied to a fixed location: communication is no longer from one place to another, but rather between people.[111] Mobile communications allow multiple connections with peers, with other groups engaged in the same cause, with the media, and with society at large. In that sense, civil activists need the internet to enable these unlimited local to global connections.[112]

Mobile phones and the internet are new tools that support the development and internationalization of civil society movements. They allow one to coordinate actions and communications without a large administrative body. For instance, mobile communication can allow the redistribution of real-time scarce resources, or the sharing of information in real time, for instance the location of the police in order to continue a protest elsewhere.[113]

As discussed in this first part, ICTs allowed the emergence of a new public sphere and new patterns of interactions that are first embraced by the Net generation, and allowed the emergence of new forms of activism. The generalization of ICTs is therefore highly connected to the development of the nonprofit sector, and more particularly to the increase in the number of civil society organizations in the world.

4. Concluding remarks

Internet and mobile-based communication campaigns enable causes to become more visible and social ties between potential participants to be created away from traditional and sometimes censored media channels. Contrary to the situation in the past, social media and web 2.0 enable the production of new content directly by the public.

In the passionate debate about the use of ICTs by civil society that opposes enthusiasts and skeptics, some elements are not disputed. First, new ICTs reduce the cost and increase the speed, ease, and reach of information exchange. Thanks to the internet for instance, access to knowledge, opinions, and ideas is at a level never seen before. Thanks to new ICTs, new organizational models (more open, horizontal, fluid, and dynamic) and civil interaction will generate more engagement and debate, and therefore strengthen civil society.[114] Technology is an ongoing process, with one innovation triggering another. The questions raised by the technological race are numerous, but this is likely to be only the beginning of the discussion about the role of technology in society and on the international stage.

New ICTs are not immune to criticism. In some cases, the interconnections between cultures and nations can lead to an intensification of conflicts. Indeed, fast-paced and permanent communication often means that it lacks depth and might not be well researched, or the sources might not be doubled-checked. Major news agencies find it more and more difficult to find the original source of a buzz started on Twitter. Sometimes, large numbers of followers or tweeters make the story plausible. The public debates around disinformation and the fake news allegations from many states including Russia, the European Union, and the USA show how serious this issue is.

Disinformation and false news are serious issues, and the role of the new ICTs is far from neutral in their propagation. According to a 2018 MIT study, fake news is 70% more likely to be retweeted than true stories. This implies for instance that false news reaches an audience of 1,500 Twitter users six times faster, results in an unbroken retweet chain[115] depth of ten levels about 20 times faster, and is retweeted by unique Twitter users more widely than real facts. Surprisingly, this is not due to computer programs that automatically retweet false news, but mainly due to retweets by individuals.[116] These results not only show worrying aspects of a contemporary online communication ecosystem,[117] but also illustrate quite well how integrated, self-sufficient, and pervasive the new global public space is. False news can circulate almost in a silo on social media channels independently from the rest of the media ecosystem, and without any root in a traditional media outlet or facts. This can have a substantial impact on political campaigns and leaders, pushing forward new agenda items to the top of the list of concerns of political and business leaders.

As discussed in this chapter, ICTs have played a key role in providing civil society actors with new means of information, collaboration, and participation. The following chapters of this book will focus on some of the current and emerging technologies used by civil society.

Notes

1 Issawi, C., 1980. Europe, the middle east and the shift in power: Reflections on a theme by Marshall Hodgson. *Comparative Studies in Society and History*, 22(4), p. 487.
2 Mayer-Schoenberger, V., Cukier, K., 2013. *Big data: A revolution that will transform how we live, work, and think*. New York, NY: Houghton Mifflin Harcourt, p. 171.
3 "ICTs" and "digital technologies" are used interchangeably to describe all ICTs that are connected to the internet or came after its invention.
4 Edwards, M., 2014. *Civil society*. Cambridge, UK: Polity Press, p. viii.
5 See Aristotle, 1970. *Physics, books I – II*. Translated with introduction and notes by William Charlton. Oxford, UK: Clarendon Press.
6 See Von Humboldt, A., Bonpland, A., 1807. *Essay on the geography of plants*. Chicago: University of Chicago Press.
7 See Malthus, R.T., 1826. *An essay on the principle of population, or a view of its past and present effects on human happiness: With an inquiry into our prospects respecting the future removal or mitigation of the evils which it occasions*. London, UK: John Murray.
8 See Darwin, C., 1859. *On the origin of species by means of natural selection, or, the preservation of favoured races in the struggle for life*. London, UK: John Murray.
9 See Haeckel, E., 1866. *Generelle morphologie der organismen*. Berlin: G. Reimer.
10 See Marsh, G.P., 1864. *Man and nature: Or, physical geography as modified by human action*. Seattle: University of Washington Press.
11 Commission on Global Governance, 1994. *Our global neighbourhood*. Oxford, UK: Oxford University Press., p. 4.
12 World Commission on Environment and Development, 1987. *Our common future*. Oxford, UK: Oxford University Press. www.un-documents.net/our-common-future. pdf
13 International Union for Conservation of Nature and Natural Resources, 1980. *World conservation strategy: Living resource conservation for sustainable development*. Gland, Switzerland: IUCN.
14 Ibid.

15 World Commission on Environment and Development, 1987. *Our common future*. Oxford, UK: Oxford University Press. www.un-documents.net/our-common-future.pdf

16 UN, 1987. *Resolution of the united nations general assembly*. New York, NY: United Nations. https://digitallibrary.un.org/record/153026/files/A_RES_42_187-EN.pdf

17 UN, 1992. *The rio declaration on environment and development*. New York, NY: United Nations. www.un.org/documents/ga/conf151/aconf15126-1annex1.htm

18 Ibid.

19 UN, 1992. *The agenda 21*. New York, NY: United Nations. https://sustainabledevelopment.un.org/content/documents/Agenda21.pdf

20 Ibid.

21 Ibid.

22 UN, 1992. *Non-legally binding authoritative statement of principles for a global consensus on the management, conservation and sustainable development of all types of forests*. New York, NY: United Nations. www.un.org/documents/ga/conf151/aconf15126-3annex3.htm

23 Ibid.

24 UN, 1992. *United Nations framework convention on climate change*. New York, NY: United Nations. https://unfccc.int/resource/docs/convkp/conveng.pdf

25 UN, 1992. *Convention on biological diversity*. New York, NY: United Nations. www.cbd.int/doc/legal/cbd-en.pdf

26 UN, 1992. *United Nations convention to combat desertification in those countries experiencing serious drought and/or desertification, particularly in Africa*. New York, NY: United Nations. www.unccd.int/sites/default/files/relevant-links/2017-01/UNCCD_Convention_ENG_0.pdf

27 UN, 2018. *Major groups and other stakeholders*. Sustainable Development Knowledge Platform. https://sustainabledevelopment.un.org/mgos

28 UN, 2000. *Millennium declaration*. New York, NY: United Nations. www.un.org/millennium/declaration/ares552e.htm

29 UN, 2000. *Millennium development goals*. New York, NY: United Nations. www.un.org/millenniumgoals/

30 Ibid.

31 Millennium Forum, 2000. *We the peoples millennium forum declaration and agenda for action*. www.i-p-o.org/millennium_forum.htm

32 Ibid.

33 UN, 2002. *Plan of implementation of the world summit on sustainable development*. New York, NY: United Nations. www.un.org/ga/search/view_doc.asp?symbol=A/CONF.199/L.7&Lang=E

34 Ibid, Paragraph. 168.

35 Ibid, Paragraph. 170.

36 UN, 2018. *Major groups and other stakeholders*. Sustainable Development Knowledge Platform. https://sustainabledevelopment.un.org/mgos

37 United Nations, 2012. *Resolution of the UN general assembly 66/288: The future we want*. New York, NY: United Nations, p. 8.

38 United Nations, 2013. *Resolution of the UN general assembly 67/290: Format and organizational aspects of the high-level political forum on sustainable development*. New York, NY: United Nations, Paragraph. 15. www.un.org/ga/search/view_doc.asp?symbol=A/RES/67/290&Lang=E

39 SDG Target 17.16 Enhance the global partnership for sustainable development, complemented by multi-stakeholder partnerships that mobilize and share knowledge, expertise, technology and financial resources, to support the achievement of the sustainable development goals in all countries, in particular developing countries.

40 17.17 Encourage and promote effective public, public-private and civil society part-nerships, building on the experience and resourcing strategies of partnerships.
41 17.6 Enhance North-South, South-South and triangular regional and international cooperation on and access to science, technology and innovation and enhance knowl-edge sharing on mutually agreed terms, including through improved coordination among existing mechanisms, in particular at the United Nations level, and through a global technology facilitation mechanism.
42 17.7 Promote the development, transfer, dissemination and diffusion of environmen-tally sound technologies to developing countries on favourable terms, including on concessional and preferential terms, as mutually agreed.
43 17.8 Fully operationalize the technology bank and science, technology and innovation capacity-building mechanism for least developed countries by 2017 and enhance the use of enabling technology, in particular information and communications technology.
44 UN, 2018. *Major groups and other stakeholders*. Sustainable Development Knowl-edge Platform. https://sustainabledevelopment.un.org/mgos
45 Gemmill, B., Bamidele-Izu, A., 2002. The global environmental agenda: Origins and prospects. In Esty, D.C., Ivanova, M.V. (ed.). *Global environmental governance: Options & opportunities*. New Haven, CT: Yale University Press, p. 82.
46 Kaldor, M., 2004. *Global civil society: An answer to war*. Cambridge, UK: Polity Press, p. 46.
47 Edwards, M., 2014. *Civil society*. Cambridge, UK: Polity Press, p. 3.
48 Kaldor, M., Anheier, H.K., Glasius, M., 2003. *Global civil society 2003*. Oxford, UK: Oxford University Press, p. 159.
49 Ibid.
50 Ibid, p. 35.
51 Clark, A.M., Friedman, E.J., Hochstetler, K., 1998. The sovereign limits of global civil society: A comparison of NGO participation. *UN world conferences on the environ-ment, human rights, and women. World Politics*, 6(1), p. 15.
52 Florini, A., 2000. *The third force: The rise of transnational civil society*. Washington, DC: The Carnegie Endowment for International Peace, p. 7.
53 Gaventa, J., 2001. *Global citizen action*. London, UK: Earthscan Publications, p. 276.
54 Kaldor, M., Moore, H., Selchow, S., 2012. *Global civil society 2012: Ten years of criti-cal reflection global civil society yearbook*. London, UK: Palgrave Macmillan, p. 21.
55 Kaldor, M., Anheier, H., Glasius, M., 2003. *Global civil society 2003*. Oxford, UK: Oxford University Press, p. 37.
56 Powell, F.W., 2007. *The politics of civil society: Neoliberalism or social left?* London, UK: Polity Press, p. 117.
57 Kaldor, M., 2004. *Global civil society: An answer to war*. Cambridge, UK: Polity Press, p. 105.
58 Ibid, p. 5.
59 Edwards, M., 2014. *Op Cit.*, p. 13.
60 Florini, A., 2000. *The third force: The rise of transnational civil society*. Washington, DC: The Carnegie Endowment for International Peace, p. 211.
61 Clark, A.M., Friedman E.J., Hochstetler, K., 1998. The sovereign limits of global civil society: A comparison of NGO participation in *UN world conferences on the environ-ment, human rights, and women. World Politics*, 51(1), p. 27.
62 Nowrot, K., 1999. Legal consequences of globalization: The status of non-governmental organizations under international law. *Indiana Journal of Global Legal Studies*, 6, p. 320.
63 Ohanyan, A., 2008. *NGOs, IGOs, and the network mechanisms of post-conflict global governance in microfinance*. New York, NY: Palgrave Macmillan, p. 6.
64 Kaldor, M., Anheier, H., Glasius, M., 2003. *Op Cit.*, p. 148.
65 Kaldor, M., Moore, H., Selchow, S., 2012. *Global civil society 2012*. London, UK: Palgrave Macmillan, p. 211.

66 Keck, M.E., Sikkink, K., 1998. *Activists beyond borders: Advocacy networks in international politics*. Ithaca, NY: Cornell University Press, p. 80.
67 Rugendyke, B., 2007. *NGOs as advocates for development in a globalising world*. London, UK: Routledge, p. 124.
68 Jamison, A., 2001. *The making of green knowledge: Environmental politics and cultural transformation*. Cambridge, UK: Cambridge University Press, p. 111.
69 Ibid.
70 Jamison, A., 2001. *Op Cit.*, p. 16.
71 United Nations Environment Programme, 1992. *Agenda 21*. New York, NY: United Nations.
72 Ibid.
73 Avant, D., Finnemore, M., Sell, S.K., 2010. *Who governs the globe?* Cambridge, UK: Cambridge University Press, p. 17.
74 Ibid.
75 Ruggie, J.G., 2004. Reconstituting the global public domain – Issues, actors, and practices. *European Journal of International Relations*, 10(4), p. 518.
76 Avant, D., 2010. *Op Cit.*, p. 12.
77 Ruggie, J.G., 2004. *Op Cit.*, p. 511.
78 Avant, D., 2010. *Op Cit.*, p. 14.
79 Zürn, M., 2012. Global governance and legitimacy problems. *Government and Opposition*, 39(2), p. 86.
80 IPCC Assessment Reports. www.ipcc.ch/publications_and_data/publications_and_data_reports.shtml#1
81 Edwards, M., 2014. *Op Cit.*, p. 19.
82 Peters, G.B., 2012. Information and governing: Cybernetic models of governance. In Levi-Faur, D. (ed.). *Oxford handbook of governance*. Oxford, UK: Oxford University Press, p. 40.
83 See Krisch, N., 2012. Global governance as public authority: An introduction. *International Journal of Constitutional Law*, 10, pp. 976–987.
84 Although this book focuses on non-for-profit organizations, it is important to notice here the emergence of the concept of Corporate Social Responsibility (CSR) in response to the externalities triggered by globalization and the growth of global multinational corporations.
85 Roughan, N., 2013. *Authorities: Conflicts, cooperation, and transnational legal theory*. Oxford, UK: Oxford University Press, p. 129.
86 Raustiala, K., Victor, D.G., 2004. The regime complex for plant genetic resources. *International Organization*, p. 278.
87 Vabulas, F., Snidal, D., 2013. Organizations without delegation: Informal Intergovernmental Organizations (IIGOs) and the spectrum of intergovernmental arrangements. *The Review of International Organizations*, 8(2), p. 195.
88 Kaldor, M., 2004. *Op Cit.*, p. 12.
89 Rosenau, J.N., 2003. *Distant proximities: Dynamics beyond globalization*. Princeton, NJ: Princeton University Press, p. 257.
90 Kaldor, M., 2004. *Op Cit.*, p. 8.
91 Ibid, p. 9.
92 Castells, M., 1996. *The rise of the network society: The information age: Economy, society and culture Vol. I*. Oxford, UK: Blackwell, p. 30.
93 Grinin, L., 2007. Periodization of history: A theoretic-mathematical analysis. In *History & mathematics*. Moscow, Russia: KomKniga, p. 20.
94 Castells, M., 1996. *Op Cit.*, p. 32.
95 Ibid, p. 30.
96 Ibid, p. 33.
97 Tapscott, D., 2008. *Wikinomics: How mass collaboration changes everything*. New York, NY: Portfolio and Penguin Group, p. 63.

98 Ibid, p. 131.
99 Mohr, N., 2013. *Mobile web watch 2012*. New York, NY: Accenture Publishing, p. 16.
www.accenture.com/us-en/Pages/insight-mobile-web-watch-2012-mobile-internet.
aspx [Accessed May 2013].
100 Tapscott, D., 2008. *Op Cit.*, p. 75.
101 Ibid.
102 Edwards, M., 2014. *Op Cit.*, p. 68.
103 Ibid, p. 67.
104 Tapscott, D., 2008. *Op Cit.*, p. 45.
105 Ibid, p. 58.
106 Deibert, R., 2012. The growing dark side of cyberspace (. . . and What to do about it).
The Penn State Journal of Law & International Affairs, 1(2), p. 263.
107 Castells, M., 1996. *Op Cit.*, p. 31.
108 Ruggie, J.G., 2004. *Op Cit.*, p. 519.
109 Kaldor, M., 2004. *Op Cit.*, p. 103.
110 Manuel Castells, 1998. *End of millennium, the information age: Economy, society and
culture Vol. III*. Oxford, UK: Blackwell, p. 154.
111 Miard, F., *Op Cit.*, p. 130.
112 Ibid, p. 240.
113 Castells, M., 2007. Communication, power and counter-power in the network society.
International Journal of Communication, 1, p. 241.
114 Edwards, M., 2014. *Op Cit.*, p. 81.
115 Also called Twitter's "cascade."
116 See MIT News 2018. http://news.mit.edu/2018/study-twitter-false-news-travels-faster-
true-stories-0308
117 See Deb Roy, associate professor of media arts and sciences at the MIT Media Lab
and director of the Media Lab's Laboratory for Social Machines (LSM). http://news.
mit.edu/2018/study-twitter-false-news-travels-faster-true-stories-0308

2 Websites and the advocacy role of environmental civil society organizations

1. Introduction

The internet, social media, mobiles, and other recent digital technology advances have unleashed powerful forces that shape and reshape local, national, and international life. Thanks to information and communication technologies (ICTs),[1] citizens and civil society have gained access to information in an unprecedented manner, which leads to an increasing politicization of global issues and institutions. Due to the rise of issues without borders, such as climate change, decisions taken on the international stage have a concrete impact on individuals at the local level. Governments, intergovernmental organizations, scientific bodies, civil society organizations (CSOs), and academia share a vast amount of information online thanks to their websites, collaboration platforms, and open document policies. Thanks to websites, individuals from around the world have gained new capacity to better grasp what is at stake in global negotiations. These recent technologies, combined with global issues, "produce and reproduce the sense of being in the world with others toward common good."[2]

ICTs change the way citizens, communities, and organizations can interact with each other. In a world characterized by heterarchy with overlap, multiplicity, mixed ascendancy, and divergent but coexistent patterns of relationships,[3] it seems difficult to disagree with Castells' observation that "the exercise of power relationships is decisively transformed in the new organizational and technological context derived from the rise of global digital networks of communication."[4] Civil society in particular has seen its ability to influence the international agenda grow thanks to digital technologies: individuals and CSOs can discuss current issues related to environmental governance and collaborate in new online spaces of interaction outside the control and monitoring of traditional forms of authority such as states and intergovernmental organizations. The Arab Spring, Occupy Wall Street, and Hong Kong's Umbrella Revolution are some well-known examples of civil movements that adopted ICTs to pursue their objectives. These new alternatives and strategies should allow civil society organizations to increasingly have their interests articulated in global rule formation institutions.[5]

Although CSOs are often confined to observer status, their role in global environmental governance is crucial at many levels: they develop and implement field projects, produce and distribute scientific knowledge, monitor governmental actors and multinationals, advocate for new policies and make proposals, take part in international negotiations, and raise awareness about issues related to the protection of the environment. This chapter focuses on one of these roles: advocacy. Advocacy is defined as the action of publicly supporting or recommending a specific cause, policy, or idea on behalf of others.[6] The objective of advocacy is to influence decision makers and the political agenda at international, national, and local levels on behalf of other stakeholders. This is one of the main activities of civil society – for two reasons. First, as mentioned previously, CSOs are rarely granted official status to participate in formal policy- and rule-making processes. Second, CSOs represent (or claim to represent) a group of stakeholders with specific interests. In terms of representation, social media is an ideal platform, since it allows direct interaction with the stakeholders the organization aims to represent. Furthermore, due to the complexity and wide range of actors and interests, advocacy also plays a major role in global environmental governance. All stakeholders involved advance their interests and pursue some of their objectives through advocacy. As an example, UN Environment mentions it as part of its main mission, which is to "serve as an authoritative advocate for the global environment."[7]

It is undeniable that ICTs provide additional means for CSOs to influence the international agenda thanks to wider access to information, new online channels to raise awareness, and new modes of interaction with decision makers such as states and intergovernmental organizations. In a world where multilateralism is increasingly decried, and where some states call for more protectionism, CSOs represent an alternative global governance path, open to a wide variety of stake-holders on the international stage. By positioning themselves as the voice of the people, many CSOs intend to represent populations and communities on the inter-national stage, the United Nations Environmental Assembly (UNEA) for exam-ple. By doing so, they support a distinct set of interests that sometimes contradicts states and intergovernmental organizations' agenda. To bring their voices to the international arena, CSOs need to use technological means to inform the stake-holders they aim to represent, as well as consult and interact with them.

This chapter will analyze the use of websites by CSOs accredited to UNEA. Since these CSOs made the effort to become accredited to this global govern-ance mechanism, it implies that they aim at influencing the debates taking place during the Assembly and their outcome. This aspect of the work of civil society organizations – setting the international agenda and influencing governmental decisions – allows them to represent the alternative set of interests mentioned previously and produce an impact on the international stage. This chapter does not aim at evaluating the impact of CSOs on the outcome of UNEA meetings, but rather to examine how they use websites to advocate. Indeed, websites offer CSOs unprecedented opportunities to inform, consult, and mobilize their audience. In that sense, websites are one type of technology used to influence international debates. Social media platforms are another set of tools that support CSOs in their advocacy strategies, which is discussed in the following chapter.

Websites are the primary instrument used by CSOs to acquire visibility among the general public, provide information about their activities, and convey their message to influence the international agenda. In other words, websites are instruments of advocacy, in the sense that they allow wide dissemination of information as well as interaction with one's audience. In that context, websites offer unprecedented opportunities to raise awareness about environmental issues and put pressure on state representatives. Providing broader access to information, consulting with one's base audience, and mobilizing stakeholders correspond to the three key communicative functions for CSO advocacy work:[8] "information" (providing information relevant to one's audience), "community" (interacting with one's audience), and "action" (calling for action).[9] These three functions were also identified by the OECD in 2001 as three forms of participation: "information" (access to useful and accurate information), "consultation" (possibility to co-design documents), and "active participation" (taking part in governance processes).[10] This means that they also support Sustainable Development Goal (SDG) 16, "Peace, justice and strong institutions,"[11] focusing on good governance and in particular SDG Target 16.6, "Develop effective, accountable and transparent institutions at all levels,"[12] and SDG Goal 17, entitled "Revitalize the global partnership for sustainable development."[13] Their advocacy strategies also allow CSOs to be better represented on the international stage, and therefore contribute to the emergence of a global partnership and more transparent institutions.

This chapter will first introduce briefly the methodology used, and then discuss how 15 CSOs accredited to UNEA use their websites to provide information, consult with their audiences, and mobilize them to support their advocacy strategies and promote their cause on the international stage.

2. Methodology

Since this chapter aims to examine how some CSOs in global environmental governance use their websites to support their advocacy strategies, the choice was made to focus on the 322 non-governmental organizations (NGOs)[14] accredited to UNEA. This assembly was created at the 2012 United Nations Conference on Sustainable Development, often referred to as the Rio+20 Conference. UNEA's Member states adopt resolutions and calls to action in order to coordinate intergovernmental initiatives related to the environment. They meet every two years to address global environmental challenges, set priorities for new environmental policies, and adopt international environmental law.[15]

Although UNEA Member state representatives are the only ones allowed to vote, major non-governmental groups and stakeholders are granted access and observer status to UNEA. More precisely, the Assembly has accredited organizations from the following groups: NGOs, the scientific and technological community, business and industry, children and youth, women, indigenous people and their communities, farmers, local authorities, workers, and trade unions.[16] Observer status allows accredited organizations to participate as observers in the Plenary, the Committee of the Whole, and the Ministerial Consultation's discussions. They can also circulate written statements to state representatives through the UN Environment Secretariat. Finally, they can make oral statements during the discussions of the UN Environment Assembly.[17] To be accredited to the Assembly, non-governmental organizations must show proof that they adhere to the following criteria: engagement in the field of the environment, international scope of work, not-for-profit, officially registered as a legal entity in at least one country, and at least two years of existence.[18]

This chapter will focus on the 15 organizations with the biggest audience on Facebook and Twitter platforms as identified in the following chapter about social media. The decision was made to offer a consistent analysis of the use of the two most common types of digital advocacy tools, namely websites and social media platforms, by these organizations on the international stage. These organizations are: World Wide Fund for Nature (WWF),[19] Greenpeace International,[20] The Nature Conservancy (TNC),[21] African Wildlife Foundation (AWF),[22] Sierra Club, Natural Resources Defense Council (NRDC),[23] Ocean Conservancy,[24] International Fund for Animal Welfare (IFAW),[25] Defenders of Wildlife (DOW),[26] World Animal Protection,[27] the Turkish Foundation for Combating Soil Erosion for Reforestation and the Protection of Natural Habitats (TEMA),[28] Conservation International (CI),[29] Earth Day Network (EDN),[30] Environmental Defense Fund (EDF),[31] and BirdLife International.[32]

To evaluate how CSOs use their websites to pursue their advocacy strategies, a web content analysis of each environmental organization's global website[33] was performed. The idea was to propose a methodology that is fairly straightforward and accessible to everyone, without the need to use pricy software tools, and without the need to have access to internal data.[34] The content of these websites was analyzed in August 2018. Five dimensions were examined. The first three dimensions correspond to the main three aspects of advocacy and participation as described previously: "information," "consultation," and "action." The other two dimensions are specific to websites and focus on the instrument itself : "website usability" and "website maturity." The choice was made to give the same weight to the five dimensions since they can be considered of equal importance. Each dimension is weighted at 20% and consists of a cluster of ten items with a similar objective or function. Each item corresponds to an element the website should provide, either in terms of technology or in terms of content, in order to support some aspect of participation, e.g. "Publications and background documentation" or "Volunteer for the organization." An item when appearing on the website is rated "1" and when missing is rated "0."[35] A total of 50 items was analyzed. Each item was weighted at 2% to reach a total of 100%. Each item corresponds to either one or several elements that must figure on the website of these 15 organizations. This means that at least one of these elements must be present, as they represent the same type of function.

The first dimension, "information" (ten items), refers to the extent of the information an organization provides. This dimension is quite broad and describes how ICTs help CSOs provide access to relevant and useful information. It includes items such as "e-Newsletter or mailing list" and "History, background information about the organization."

The second dimension, "consultation," corresponds to the various options that the organization offers on its website to receive feedback from its audience. This is particularly crucial for CSOs that represent the voice of local communities or individuals, or that support a specific cause. It includes "Commenting on news items/blog posts," "Contribution etiquette/community policy/terms of use," and "Make a suggestion/proposal/send feedback," among other items.

The third dimension, "action," encompasses all the options offered to stakeholders to contribute to the work of the organization. Hence, it is not limited to taking part in decision-making processes. Indeed, this dimension includes a broader understanding of action, and embeds elements such as "Pledge/adopt a new behavior," "Volunteer for the organization," and "Become a member online/donate online." In this dimension, donating and fundraising are separate. Indeed, to fundraise requires more involvement from the audience, and at the same time more resources from the organization. On the other hand, while donating online and becoming a member aim principally at supporting the organization financially, they require less time and effort from the audience and fewer resources from the organization.

The fourth dimension, "website usability," refers to the website's ease-of-use by the audience.[36] This dimension is important since it allows a wide variety of users to make the most out of the website. It focuses on the more technical aspects of digital participation, including "Search engine for publications and news items" and "Site map or A to Z Index (Alphabetical Index)," among others. The choice was made to also include here some measures of web traffic. Indeed, it is important to evaluate how a website is in effect used by the audience, and not only focus on what an organization offers. Since the web content analysis is based on identifying whether an element is present or not on the website, the analysis of web traffic must follow the same rule. A benchmark was therefore defined to identify when a website was considered "in effect used by the audience" or not. Since there is no general consensus on how to evaluate web traffic for each organization (due to its specificities such as type and location of audience, size, objectives, and sector), the choice was made to use the median of all 15 organizations' web traffic as a benchmark. Three aspects of web traffic were measured in this dimension: daily pageviews per visitor (over a 30-day period), Alexa traffic ranking (over a 30-day period), and number of websites linking in (over a 30-day period). The first element of traffic shows how many pages a web user consults on average; this provides an indication of the relevance of the content. The second element is an estimate of the website's popularity developed by Alexa. It combines the average number of daily visitors and pageviews over the last three months. Since it is a ranking, the lowest result is the best; i.e. the most popular website in terms of traffic and pageviews in the world is ranked 1. The third element shows how many websites provide links to the organization's website. This shows how well known and recognized by others the website is. In other words, these three elements of traffic provide an indication of the content viewed, number of visitors, and recognition.

The last dimension, "website maturity," focuses on the degree of sophistication of websites.[37] It provides an indication of how advanced the website is in terms of features and content. It also allows an evaluation of how many resources are dedicated to digital participation, and therefore how important this channel of communication is for the organization. It comprises items such as "Secure servers (https://…)," "Login with passwords to access dedicated part of website," and "Multimedia (video, audio, images)."

In order to extract some general trends from this sample of 15 organizations, the means (averages) for each item and dimension were calculated and are shown in Tables 2.1–2.5. Additionally, the scores for all five dimensions are averaged in Table 2.1 and show the final result for each organization. The next sections will discuss the five dimensions mentioned previously: the three main uses of websites to support CSOs' advocacy strategies – information, consultation, and action – as well as the levels of website usability and website maturity.

3. Providing access to information

Access to information is crucial when it comes to raising awareness about specific issues and setting the agenda on the international stage. It is also highly important for CSOs that aim to be as transparent and accountable as possible. By providing large amounts of information to the public,[38] they often intend to bridge the gap between decision makers in international arenas and citizens at the local level. This dimension shows how the 15 CSOs from this sample use their websites to provide broad access to information, which supports their advocacy strategies on the international stage. Table 2.1 shows the mean (average) of all 15 organizations' scores for the "information" dimension.

This first dimension (see Table 2.1) is predominantly about content (nine items out of ten), implying that it is less costly to implement. It is by far the most

Table 2.1 Information dimension: average of organization scores (%)

1. INFORMATION – ten items (20%)	Average
CONTENT: Contact details: telephone/online form/address/email	100%
CONTENT: History/background information	100%
CONTENT: Mission statement/vision	100%
CONTENT: Main fields and topics of work/projects/activities	100%
CONTENT: Top management organigram/governance	87%
CONTENT: Newsroom/news stories/press release/blog	100%
CONTENT: Publications/knowledge content	100%
CONTENT: Budget/annual reports	93%
CONTENT: Audit/external monitoring reports	73%
TECHNOLOGY: e-Newsletter/mailing list	100%
TOTAL	**95%**

developed, with an average of 95% for all organizations. Table 2.6 shows the results for each organization individually.

This result shows that on average the 15 organizations from this sample have developed and implemented 95% of the access to information items measured here. It indicates that CSOs use websites extensively to pursue this aspect of advocacy. It must be noted that most of these 15 organizations are global organizations, which means that their audiences are spread out geographically. Without a website, these organizations could not have the same outreach and the same impact on their audiences. This positive result is not surprising, since this category contains some of the most "basic" elements of a website such as "Contact details: telephone/online form/address/emails," "History/background information," "Mission statement/vision," "Newsroom/news stories/press release/blog," and "Publications/knowledge content." All organizations provide background information about the topics related to their work, not only by producing news items and blog posts about their latest activities, but also by producing specific knowledge to ensure that their audiences value the cause they are defending and understand what is at stake. These are essential elements to raising awareness about nature conservation, which is a complex issue requiring some basic knowledge in order to better grasp current issues such as deforestation and climate change.

Interestingly, most organizations also show online their "Top management organigram and governance" (87%). These are important elements of transparency and accountability, along with "Audit/external monitoring reports" (73%) and "Budget/annual reports" (93%). This information allows all stakeholders, including individuals, to have access to relevant information and decreases the traditional information asymmetry between organizations and their audiences. The least represented item of this dimension, "Audit/external monitoring reports" (73%), is due to the fact that in some cases the auditing reports are available only on request, as is the case for WWF for instance.[39] In other words, this is not necessarily an indication of a lack of transparency, but could be explained by a management choice to provide this information only on request and not directly on the website.

4. Consulting with one's audience

Since all organizations work in the field of nature conservation, they represent a community of activists, conservationists, groups, individuals, and sometimes organizations that defend the same cause. Openness and the possibility for mass collaboration and participation are expected to gradually change organizations and harness collective intelligence, building on the knowledge, experience, and competence of various actors.[40] CSOs develop advocacy strategies to raise awareness among populations and put pressure on state representations. In that context, consultation is crucial to ensure that these organizations (1) understand their audiences, (2) benefit from the knowledge and experience of their audiences, and (3) represent their audiences effectively on an international stage such as UNEA. This is precisely what this dimension aims to assess through analysis of the websites of the 15 organizations.

This dimension contains more items related to technology (eight) and is more resource intensive. It shows averages much lower than for the previous dimension. Indeed, with only 24% of all consultation items implemented here, this indicates that these 15 CSOs have not developed many consultation features on their websites. This could appear to be a surprising result. Indeed, most of these organizations do not allow for comments on their websites. Although comments increase the participation of the audience and enable the organization to benefit from the knowledge and experience coming from external web users, they also provide a space for negativity and vulgarity. When everyone is allowed to share their point of view, it requires the organizations to monitor these comments, allowing some, refusing others, and explaining why. This is highly resource intensive, and for many organizations, this may be why the choice was made not to allow any commenting. As shown in Table 2.2, only 40% of all organizations allow "Commenting on news items/blog posts," and only 7% of them allow "Commenting on publications/knowledge content." None of them allow "Commenting on governance/reporting/monitoring documents" or "Commenting on multimedia: videos/audios." Furthermore, very few organizations allow web users to "Report a fraud/incident (other than website)" (7%), "Make a suggestion/proposal/send feedback" (13%), or consult through "Web surveys" (7%). Lastly, few organizations have a web forum for stakeholders to debate issues related to the cause the organization supports (7%).

These limited results are not only due to the fact that commenting requires substantial resources. They are also due to the emergence of social media platforms. All of these organizations are social media champions[41], which means they have shifted the consultation dimension to these platforms, where it is not only less resource intensive, but also more concentrated on a few channels, with users clearly identified. Therefore, the results of this dimension and the following one should be evaluated along with CSOs' social media presence, which is strong and active.

Table 2.2 Consultation dimension: average of organization scores (%)

2. CONSULTATION – ten items (20%)	Average
CONTENT: Contribution etiquette/community policy/terms of use	67%
CONTENT: Privacy policy	93%
TECHNOLOGY: Commenting on news items/blog posts	40%
TECHNOLOGY: Commenting on publications/knowledge content	7%
TECHNOLOGY: Commenting on governance/reporting/monitoring documents	0%
TECHNOLOGY: Commenting on multimedia: videos/audios	0%
TECHNOLOGY: Web surveys	7%
TECHNOLOGY: Make a suggestion/proposal/send feedback	13%
TECHNOLOGY: Report a fraud/incident (other than website)	7%
TECHNOLOGY: Web forum for stakeholders to discuss	7%
TOTAL	**24%**

The highest result of this dimension corresponds to an important aspect of consultation: how personal data are used and collected. A total of 93% of these organizations include a privacy policy on their websites, showing that these global CSOs take confidentiality and privacy seriously. Moreover, 67% of these organizations show "Contribution etiquette/community policy/terms of use" on their websites. This is quite a high result and indicates that in addition to realizing that ICTs can empower transparency and consultation, these organizations also recognize the necessity of having clear rules and a code of conduct on how to use their websites.

5. Action: mobilizing one's audience

The third dimension focuses on action and the mobilization of the audience to support a specific cause and the activities of the organization. One of the most important impacts that ICTs may have on advocacy strategies is to increase the capacity of organizations to mobilize their audience to act in a certain way, including to provide additional funding sources, adopt a more environmentally friendly behavior, take part in cleaning beaches, and share information with their own network to raise awareness about a cause, as shown in Table 2.3.

This third dimension shows similar results to the previous one, indicating that CSOs have developed on average only 25% of the action items examined here. It is also quite a technology-intensive dimension with six items related to technology. With on average 47% of fundraising features developed on their websites, these organizations tend to offer a wide range of possibilities to financially support their work. As mentioned previously, fundraising is dissociated from donating online, since it requires much more active participation and involvement than simply donating money. Moreover, all organizations offer the possibility to become a member or donate online. This is a constant feature on all websites but covers a wide range of different options, from donating money to donating a car, from adopting an animal to donating in honor of someone, from estate donations to corporate donations. It is also an important part of the work of the organizations that depend on these donations to complete their work in the field.

Table 2.3 Action dimension: average of organization scores (%)

3. ACTION – ten items (20%)	Average
CONTENT: Fundraise (organize a funding campaign)	47%
CONTENT: Volunteer for the organization	40%
CONTENT: Pledge/adopt a new behavior	60%
CONTENT: Take part in observation/citizen science	7%
TECHNOLOGY: Share on social media/show support/ambassador	93%
TECHNOLOGY: Contribute to the blog/news stories	7%
TECHNOLOGY: Sign a petition	80%
TECHNOLOGY: Join a group/community online	13%
TECHNOLOGY: Become a member online/donate online (support the org)	100%
TECHNOLOGY: Take part in governance processes/decision-making processes	0%
TOTAL	**25%**

These organizations have also extensively developed the possibility to "Sign a petition" (80%), which here as well corresponds to a large range of different options, from sending a pre-filled form to a political or business leader to adding one's name to a list. Petitions could be identified as consultation, but since they require a more active role toward the outside world than just sending a message dedicated to the organization and its web users, the choice was made to include petitions as an action.

Surprisingly, these organizations have not developed a great deal of content on their websites to support volunteering. Here, volunteering is understood to be providing work for free; it does not include internships within the organization, which correspond to a first job position after university. Volunteering encompasses all activities that individuals can perform to support the cause of the organization. A good example is cleaning the oceanfront and beaches of plastic waste for instance.

These organizations also do not offer the possibility to contribute content to the blog and to the website (7%). This percentage is quite low, even though it could increase the loyalty between the organization and its audience. Particularly for a blog, it is common practice to open it to a wider audience, allowing one, as with websites, to clearly distinguish between the content of the organization and the content that could be produced by an external audience. Here as well, the fact that most of these organizations focus interaction on social media platforms probably explains this result. Nevertheless, it could be an added value in many cases.

Another limited result is "Take part in observation/citizen science," with only 7%. This does not mean that citizen science is not utilized by these organizations, but rather that it is done through other platforms and dedicated mobile apps. The chapters on AI and big data show how citizen science is actively used by most CSOs in the environmental sector. However, few organizations have dedicated a section of their websites to these activities.

Another limited result is seen for "Join a group/community online" (13%). Although it may seem similar to the "Web forum for stakeholders to discuss" item of the previous dimension, this item refers to when a website allows web users to join online a specific community or group of conservationists who meet in the real world to undertake some type of activity to support the cause of the organization. These could include local chapters and special-interest groups who meet and gather to organize communication campaigns, organize fundraising activities, and support field projects.

A more positive result concerns "Pledge/adopt a new behavior," with on average 60% of the organizations having developed this item on their websites. This indicates another aspect of action: how the audience can support the cause of the organization by acting in a certain manner. This covers a large array of activities such as refusing plastic straws, eating less meat, or pledging to take part in a protest against a particular issue or in favor of a specific cause.

The highest result is "Share on social media/show support/ambassador" (93%), which confirms what was mentioned previously: these 15 organizations are very present on social media, and therefore use their websites first and foremost to provide access to information. Initiatives involving consultation and action benefit from the power of interaction inherent in social media platforms.

If none of these organizations allow users to "Take part in governance processes/decision-making processes," it is not only due to their governance structures not necessarily allowing the audience to take an active role in the governance process. It could be argued that since they support nature rather than representing a community or group of individuals, they do not need to have an open governance process. However, since the consultation dimension also showed low results, it can further be argued that consultation and participation in the future of the organization are achieved through social media platforms, where the organizations can perform surveys, solicit ideas and comments, and then include these comments in their governance processes. This is to be associated with the high result of social media features on their websites.

Moreover, if none of these organizations allow users to "Take part in governance processes/decision-making processes," it is also partly due to security and privacy reasons. Indeed, it remains highly complicated to ensure that a stakeholder makes an e-decision without any external influence. When a decision is made in a meeting room, influences and power plays can be numerous and evolving. However, physical presence in a conference room ensures that the decision is made according to the rules in vigor and prevents illegal pressure. Furthermore, hacking is an increasing issue. This implies that when developing new systems for decision making, CSOs need to ensure that decisions cannot be hacked or leaked. Therefore, it is understandable that these organizations have not developed this item.

6. Additional results

6.1. Website usability

Since this analysis focuses on websites as instruments to pursue advocacy strategies, it is essential to evaluate the technology itself, and not only the content and functionalities it allows a CSO to offer. For the audience to benefit fully from the information, consultation, and action features developed on websites, organizations must ensure the information is easily accessible by the vast majority of their stakeholders. This is indeed the main reason to use the internet: affordable and easy access to information on a global scale. In that sense, website usability shows how user friendly CSOs' content is, and also contributes to building CSOs as new forms of authority. Table 2.4 shows the results of the "website usability" dimension.

This dimension is also quite technology and resource intensive, with only two items about content. As shown in Table 2.4, the website usability level is fairly high (57%), but with significant differences between the results, ranging from 100% for "Direct link to official social media platforms" to 0% for "Text-only version (slow internet)/audio access (visually impaired)." It is quite surprising that global CSOs do not provide text-only accessible versions or audio access

Table 2.4 Website usability dimension: average of organization scores (%)

4. WEBSITE USABILITY – ten items (20%)	Average
CONTENT: Site map/A to Z Index	60%
CONTENT: FAQ page (frequently asked questions)	27%
TECHNOLOGY: Search engine for content on the entire website	93%
TECHNOLOGY: Search engine for publications and news items	40%
TECHNOLOGY: Mobile version of the website	93%
TECHNOLOGY: Text-only version (slow internet)/audio access (visually impaired)	0%
TECHNOLOGY: Direct link to official social media platforms	100%
TRAFFIC: Daily pageviews per visitor (content viewed)	N/A
TRAFFIC: Alexa traffic rank (number of visitors and pageviews)	N/A
TRAFFIC: Sites linking in (recognition)	N/A
TOTAL	**57%**

to the website. These are highly important for both the visually impaired and audiences located in parts of the world where internet access is limited. The well-known issue of the digital divide, which describes how internet penetration rates vary dramatically between developed and developing countries, should motivate these organizations to provide these alternative versions in order to provide access to a larger audience. This is particularly critical when modern websites make extensive use of large pictures that require a fast internet connection.

Equally surprising is the fact that "Site map/A to Z Index" shows a result of 60%, when this is a very simple and highly practical feature to implement. Here as well, the design of modern websites, with one-page design for instance, tends to omit this element. It should not be taken for granted that all web users are well experienced. The digital divide can also apply to the age gap between younger and older generations, the latter requiring more indications to be able to navigate on large websites such as those of these 15 organizations. It can at times be confusing to find the right information, especially when the search functionality that tends to replace the site map does not perform exactly the same function. Indeed, when searching for a title or a section with the search functionality, web users end up with a large number of references that include articles, news items, and sections of the website.

It is also surprising that FAQ pages (frequently asked questions) are not more developed (27%). They provide good support for web users when consulting the website for the first time, in particular when it comes to understanding the "action" features of the website. However, here as well, social media platforms, which allow direct interaction with the organization, tend to replace this functionality.

Most organizations have a search engine (93%) to help users find specific elements on their websites. This is highly useful for large websites. However, the results often show all together a large number of news items, documents, and sections of the website, and this can be confusing. A better categorization of these results would be more useful, especially when the website does not offer a search engine dedicated to news items and publications, which is the case for most websites.

Most organizations provide a mobile version of their website. This is highly useful since an increasingly large majority of users browse the web on their tablets or mobile devices. In addition, all organizations provide a direct link to their official social media platforms. This is different from the item "Share on social media/show support/ambassador" in the previous dimension. Here what is measured is not the possibility to share the website content on one's personal social media platform, but rather to access the official social media platform of the organization.

In terms of traffic, the results do not show here, given that the methodology applied is based on the median. Indeed, the benchmark used is the median, which implies that the average result will be 50%, and this does not represent any reality. In other words, the traffic results are relevant only for each organization, and not as an average for the sample. They will be included in Table 2.7 showing the results for each organization individually.

6.2. Website maturity

The last dimension includes a range of items that aim to show the level of sophistication of the websites analyzed. It also shows how many resources these organizations have dedicated to websites as channels of communication, and therefore how important they are for them. It is essential to remember that these 15 organizations are accredited to UNEA, which implies that they aim to raise awareness about issues and achieve momentum in order to influence the international agenda and put pressure on state representatives and decision makers. In that sense, websites are the windows to the world of these organizations, and therefore influence how they are perceived by various stakeholders, including governmental actors, their audience, and other partner organizations. Hence, website maturity also participates in building them as new forms of authority. Table 2.5 shows the results for the website maturity dimension.

This dimension is fairly well developed at 69%, although it contains six items related to technology. A first surprising result is the fact that not all websites offer at least two languages. The websites consulted here are global websites, which would suggest that they offer not just English as a language. However, because all of these organizations (except TEMA) are global, they have multiple websites for distinct regions and countries, and therefore offer content in different languages on these other websites, which were not taken into consideration here.

In terms of volume of content, the measure proposed here corresponds to the depth of menus, "2+ level menu" (80%). If the menus contain at least two levels, meaning a section and a subsection, they are considered to be offering a substantial volume of content. This result is in fact to be expected since these websites are for global organizations supporting global causes, and hence require large volumes of content to raise awareness. Also, all of them offer at least two types of

Table 2.5 Website maturity dimension: average of organization scores (%)

5. WEBSITE MATURITY – ten items (20%)	Average
CONTENT: 2+ languages offered to view content	53%
CONTENT: 2+ level menu (as indicative of volume of content on website)	80%
CONTENT: 2+ types of media (video/audio/pictures)	100%
CONTENT: Job listing/career section	93%
TECHNOLOGY: Related micro-websites with distinct URL (campaigns/causes/etc.)	47%
TECHNOLOGY: No broken links	87%
TECHNOLOGY: Secure servers (https://…)	87%
TECHNOLOGY: Mobile apps/web tools	27%
TECHNOLOGY: Log in/create an account (Not only sign up for newsletter)	40%
TECHNOLOGY: Modern design (large photos/one page/parallax, etc.)	73%
TOTAL	**69%**

media. In fact, the majority offer video and pictures, with only a few also offering audio content.

Most of them offer a job section, which also provides an indication of the sophistication of the website, since this requires the organizations to provide application process details online, through either a dedicated web application or an external website. It is also a question of transparency and accountability for the recruitment process.

Since these organizations support global causes, some have developed micro-websites dedicated to specific communication campaigns or initiatives (47%). This indicates how many resources are dedicated to web communication, and the extent of the organizations' web presence to raise awareness.

Surprisingly, some organizations had broken links. This shows up when one aims to access a page via a link on the website and either the page is blank or it does not exist – i.e. the link does not work. This is especially surprising for global organizations, since their web presence is also an indication of their credibility, and a broken link does not convey a good image of the organization.

In terms of security, the use of "Secure servers (https://…)" (87%) has become the norm, which is highly positive since it protects the privacy and integrity of the exchanged data. In an age of cyber-attacks and cyber-threats, it is crucial for these organizations to ensure this level of security.

Few organizations have developed mobile apps or web tools. In terms of mobile apps, not only are Android and Apple products included, but also games. Web tools comprise a range of instruments, from interactive maps to data analysis instruments. Their objective is to enable the user to perform a specific type of activity online. For some of these organizations their regional or country sections have developed specific mobile apps. However, globally, few organizations have.

Another aspect of website maturity is the possibility to "Log in/create an account" (40%). This does not cover the possibility to sign up for newsletters but rather to create a profile or an account on the extranet of the organization, where web users can add their personal contact details, maybe credit-card information, or preferences in communication and publications (in terms of both topic and regularity).

Lastly, most organizations have implemented what is identified here as "Modern design" (73%), which includes large photos and parallax technology, among others. This could be perceived as quite subjective, but only the use of large frame pictures (using the breadth of the desktop or laptop computer screen) is an indication of a website that was developed in the last few years. This is a very high result, and it shows that most CSOs of this sample take web communication very seriously, dedicating a substantial amount of their financial support to web resources to raise awareness and funds.

The next section will discuss the results of each organization.

6.3. Results per organization

If on average, these 15 CSOs have developed 54% of all five dimensions, results vary greatly from one organization to the next, from the organizations that have developed the most features on their websites (70%) to the ones that have the least (36%). The results per organizations, as shown in Table 2.6, shed more light on how CSOs use their websites to advance advocacy strategies and influence the international agenda.

First, Table 2.6 shows no specific differences between causes (ocean, animal protection, environment in general), type of organization (project, advocacy, hybrid), size, or location. Also, it shows that eight organizations out of 15 are in the range of ±10% of the average 54%. This indicates a certain homogeneity in terms of results in more than half of these organizations. TEMA is the organization showing the lowest results, and this can be explained by multiple factors. First, it is the only national organization, with its main communication channels in Turkish. This means that the English version of the website analyzed here is probably not as developed as the Turkish version. Second, the low result of the "consultation" dimension (0%) must be linked to the fairly low results of "website usability" and "website maturity" since the English version of the website is the least advanced one in terms of design and technology. But once more, this can be explained by its national audience.

The most developed dimension among all organizations is "information" (95%). Some organizations have all items measured here, while others lack some of them, in particular relating to external reviews and auditing reports, as discussed previously. In this aspect, Sierra Club performs less well than others (80%). This is due to the fact that the Sierra Club Foundation is the fiscal sponsor of Sierra Club's environmental programs.[42] This implies that all financials and governance documentation are provided on the website of the Foundation and not on the website of Sierra Club. However, in general, this is the dimension where all organizations excel; this is understandable since it is the primary purpose of a website: to communicate information. It is also the least resource- and technology-intensive dimension, and therefore the easiest to implement.

Table 2.6 Scores of the five dimensions by organization

	Information	Consultation	Action	Website usability	Website maturity	Total
The Nature Conservancy	100%	50%	30%	70%	100%	70%
Greenpeace International	90%	50%	30%	80%	70%	64%
Sierra Club	80%	40%	40%	70%	90%	64%
BirdLife International	100%	20%	40%	50%	90%	60%
World Wide Fund for Nature	90%	20%	30%	80%	70%	58%
Defenders of Wildlife	100%	30%	20%	70%	70%	58%
Environmental Defense	100%	20%	10%	70%	70%	54%
African Wildlife Foundation	100%	20%	20%	60%	70%	54%
Conservation International	100%	30%	30%	40%	60%	52%
Natural Resources Defense Council	100%	10%	20%	60%	70%	52%
Earth Day Network	100%	20%	30%	50%	60%	52%
International Fund for Animal Welfare	100%	30%	10%	60%	50%	50%
Ocean Conservancy	100%	10%	10%	40%	70%	46%
World Animal Protection	90%	10%	20%	40%	50%	42%
TEMA	80%	0%	40%	20%	40%	36%
Average	**95%**	**24%**	**25%**	**57%**	**69%**	**54%**

The second dimension, "consultation" (24%), shows similar results to the third dimension, "action" (25%). In terms of "consultation," there is a high diversity in the results, ranging from 50% to 0%. The commenting feature is missing on most websites, along with web forums and web surveys. This is largely explained by the wide use of social media platforms, as discussed previously. Indeed, these 15 organizations were chosen based on their performance on Twitter and Facebook. Moreover, some of these organizations (WWF for instance) have distinct websites for their regional and country branches. An analysis of these websites could show different results, since in this case, the international website focuses more on providing information, whereas the regional or national websites are more connected to their local audiences, which could result in stronger results in terms of "consultation." Greenpeace International (50%) is an exception here, with a commenting feature available on its website and the possibility to make a suggestion and provide feedback, which shows as a tag on most pages. The Nature Conservancy (50%) also shows above average results, with the possibility to make a suggestion and provide feedback, report a fraud or incident, and participate in web surveys. These features add real value for the web user when consulting these websites, thereby increasing the perception of transparency and interaction. Here it is important to note that all organizations have a privacy policy available on their websites, which shows that these CSOs recognize the importance of the question of privacy and confidentiality.

Concerning the third dimension "action" (25%), most organizations show limited results ranging from 40% to 10%. The low level of results here can also be explained by the same reasons mentioned for the second dimension, "consultation." Since all these organizations are accredited to UNEA, raising awareness on the international stage is crucial. Therefore, they must call out to their audiences to increase their knowledge about specific issues and motivate them to act and support the cause. This is confirmed by the presence on the websites of these organizations of the items related to raising awareness and calling on their audiences to support a cause: become a member, donate, share content on social media, pledge for a certain behavior or action. Contrary to this, other types of "action" are not encouraged by these organizations. For instance, they do not allow participation in governance processes. This is explained by the fact that opening up governance processes to external audiences can be challenging, and in any case, highly complicated to implement. This indicates that the action requested of the audience is linked mainly to raising awareness initiatives, which are reinforced on social media platforms.

The fourth dimension, "website usability" (57%), ranks quite high compared to the preceding two, but with high heterogeneity among organizations, with a range of 80% to 20%. While most organizations provide a search engine for the entire website and offer a mobile version for their websites, alternative versions for slow internet connections and for the visually impaired are missing, as discussed previously.

The last dimension, "website maturity," evaluates how sophisticated these websites are in terms of content, but mainly in terms of technology and design. This is the second-best dimension, with 69%. If most organizations use a secure server, use multimedia extensively, have no broken links, and have a modern design, few offer mobile apps and web tools for their audiences, as mentioned previously. Contrary to the other dimensions, where some organizations showed 0% of implementation, the website maturity dimension indicates that most organizations have a good level of website maturity. The organization with the lowest level of implementation is TEMA, with 40%, which once again was evaluated on the English version of its website, and therefore might not show all content and technological features that are available on the main version in Turkish. In general, it can be said that the organizations of this sample actively take into consideration this channel of communication to raise awareness and defend their cause.

6.4. Web traffic

In terms of web traffic, Table 2.7 shows detailed web traffic information from the three types of indicators used in this dimension: Alexa global traffic ranking, daily pageviews per visitor, and number of sites linking in. The results are organized based on the first indicator; Alexa global traffic ranking provides a web traffic and pageview estimation of most websites on the planet, and ranks them starting with number 1 (website with the most traffic and pageviews).

Table 2.7 Web traffic indicators for the 15 CSOs

	Alexa traffic ranking	Daily pageviews per visitor	Number of sites linking in
Greenpeace International	21,647	1.86	28,002
The Nature Conservancy	41,385	1.83	8811
Natural Resources Defense Council	46,060	1.5	8516
Sierra Club	47,911	1.82	8464
World Wide Fund for Nature	52,871	1.8	11,914
Environmental Defense	105,689	1.9	3183
Defenders of Wildlife	108,114	1.4	2285
Conservation International	151,936	1.7	2976
BirdLife International	152,571	1.9	3617
African Wildlife Foundation	178,256	3.2	1285
Earth Day Network	234,362	1.4	3998
Ocean Conservancy	246,811	1.9	1435
International Fund for Animal Welfare	279,057	2	2246
TEMA	462,685	3.2	696
World Animal Protection	639,252	1.9	338
Average	**194,802**	**1.95**	**5851**

The first indicator (Alexa traffic ranking) shows a high discrepancy in the results, and here as well there are no specific differences between types, location, size of organization, and the cause they defend. Greenpeace International ranks twice as well as the next organization, and almost 30 times better than the last one. There is moreover a small group of super-performing organizations, in terms of traffic, which rank within the top 55,000 best websites: Greenpeace International, The Nature Conservancy, Natural Resources Defense Council, Sierra Club, and the World Wide Fund for Nature. If we compare this ranking with the overall performance results for all five dimensions, it shows that four out of five are the same organizations that rank best in terms of web traffic and overall digital participation. Similarly, the five organizations with the lowest numbers in terms of traffic are the same ones that perform the least well in terms of digital participation: Earth Day Network, Ocean Conservancy, International Fund for Animal Welfare, TEMA, and World Animal Protection. The fact that web traffic and digital participation features are related should be expected. However, is it because CSOs have a lot of traffic to their websites that they develop digital participation features, or is it because they have developed digital participation features that they have more traffic? The direction of this causal relationship remains an open question, even if it would be logical to think that a more participatory website with better access to information, consultation, action, website usability, and website maturity should trigger higher levels of web traffic.

In terms of number of pageviews, the results are quite homogeneous, with about 1.95 pages daily per visitor. This indicates that almost every visitor consults two pages on the website, and does not only check the home page and bounce. This indicator is essential to assess the relevance of the content proposed on the website, and shows that most visitors find the information and features relevant enough to consult two pages. The last indicator is the number of websites linking to the CSOs' websites. This helps understand how well "connected" and "recognized" these websites are. If other websites find them relevant, credible, and legitimate as sources of information, they will link in. Here a very strong heterogeneity shows between organizations, with 28,002 websites linking to Greenpeace International's website, 11,914 to World Wide Fund for Nature's, 8,516 to Natural Resources Defense Council's, and 8,464 to Sierra Club's. These are the same top-performing organizations for overall digital participation and for web traffic – a logical outcome since the more websites linking to a CSO's website, the more traffic the latter will receive.

7. Concluding remarks

Thanks to ICTs, individuals and organizations transform a grasp of how the world works into working knowledge. This skill revolution, coined by well-known scholar James N. Rosenau, consists of an interplay between technologies, education, and experience; these three elements allow individuals to grow exponentially and enable them to catch the attention of global leaders.[43] Governments, corporations, and other collectivities have also gained new skills, but this increase pales when compared to the one at the micro-level. Indeed, the main transformation is the empowerment of the individual around the world. ICTs have accelerated the pace of the learning process: people everywhere in the world learn faster. In other words, "the world has entered the age of the person."[44] ICTs, combined with this skill revolution, have also led to an organizational revolution, where individuals and organizations come together in new ways in horizontal and vertical structures of authority.[45]

Not only do ICTs support actors in acquiring new skills, but they also contribute to the emergence of new spaces for interaction. These new structures of authority are in the early stage of development; their emergence is not likely to be signaled by some sort of founding event, but they will tend toward a developing and variable structure and nature since any social form is subject to changes in relevant contexts from one time to the next.[46] These new structures of authority correspond to a coherent configuration of organization, space, and interaction.[47] ICTs do indeed impact global institutions by giving new prominence to the communicative dimension of governance. This is crucial since relations are made primarily through the establishment of communicative interactions in a digital space.[48]

This chapter focused on a sample of 15 CSOs accredited to UNEA. Since they are actively engaged in global environmental governance, their use of websites should reflect this involvement, and support their advocacy strategies on the international stage to raise awareness about specific issues and influence the international agenda. Websites offer unprecedented opportunities to reach these two advocacy targets, their audiences, and state representatives, by providing broader access to information, consulting with their base audience, and mobilizing their stakeholders to act.

This analysis showed that CSOs have developed on average 95% of the access to information features identified here. This is indeed the first and foremost use of websites: providing information to a wide audience. However, only 24% of the consultation and 25% of the action/mobilization features were developed on average, which indicates that these CSOs do not consider their websites as primary channels to achieve these goals, despite being so essential to their advocacy objectives. This result is confirmed by their use of social media platforms, as discussed in the following chapter, where they also first and foremost inform their audience. In other words, it is safe to conclude that their advocacy strategies are mainly informative and focus on educating and raising awareness. Although websites could allow more opportunities in terms of consultation and mobilization, it is not highly surprising since social media platforms offer simple and affordable interaction and community building features. In other words, one could expect that websites are used mainly to inform, while social media platforms are dedicated to the other two aspects of advocacy. The next chapter will examine these elements.

Moreover, there is high discrepancy in the use of websites by CSOs. Some have developed many features to inform, consult with, and mobilize their audience, and built websites with high levels of usability and sophistication. Web traffic is a strong indicator of these levels of website development: CSOs with high web traffic have developed substantially more features on their websites, and conversely, organizations with less web traffic have implemented fewer features in terms of both content and technology. It seems logical to pursue the idea that a website with well-developed content, along with numerous consultation and action initiatives, accessibility and ease-of-use, and high sophistication, can only increase opportunities for transparency and interaction. At a time when information is increasingly consulted online first, and when individuals spend more and more time connected and in front of a screen, these five website features not only support CSOs' advocacy strategies, but also enhance the credibility of these organizations to be recognized as new forms of authority on the international stage. Hence, their advocacy strategies, focusing on providing information, confirm their expertise authority.

Lastly, the use of websites also sheds light on the participation capacity and strategies of CSOs on the international stage. Indeed, as mentioned previously, their advocacy strategies allow them to participate in global environmental governance by making their voices heard, and by becoming new forms of authority, and in that context support SDG Target 16.6, "Develop effective, accountable and transparent institutions at all levels,"[49] and SDG Target 16.7, "Ensure responsive, inclusive, participatory and representative decision-making at all levels." CSOs' use of websites to inform, consult with, and mobilize their audience also enables them to gain more visibility on the international stage, and doing so, also allows them to better cooperate with other actors and help achieve SDG Goal 17 entitled "Revitalize the global partnership for sustainable development."[50]

Notes

1 ICTs comprise all technologies that help gather, distribute, produce, consume, and store information, including print and broadcast media, channels of communication (satellite, cable), telecommunications (phone, web), computers, and storage devices. Definition from: Singh, J.P., 2002. Information technologies and the changing scope of global power and governance. In Rosenau, J.N. (ed.). *Information technologies and global politics: The changing scope of power and governance.* Albany: State University of New York Press, p. 2.

2 Gordon, E., Mihailidis, P., 2016. *Civic media: Technology, design, practice.* Cambridge, MA: MIT Press.

3 Brousseau, E., Marzouki, M., Méadel, C., 2012. *Governance, regulation and powers on the internet.* Cambridge, UK: Cambridge University Press, p. 18.

4 Castells, M. mentioned in: Padovani, C., Pavan, E., 2016. Global governance and ICT: Exploring online governance networks around gender and media. *Global Networks*, 16(3), p. 353. https://doi.org/10.1111/glob.12119

5 Singh, J.P., 2002. *Op Cit.*, p. 243.

6 Soanes, C., Stevenson, A., 2004. *Concise Oxford English dictionary*, 11th ed. New York, NY: Oxford University Press, p. 19.

7 UNEP, 2018. www.unenvironment.org/about-un-environment

8 Lovejoy, K., Saxton, G.D., 2012. Information, community, and action: How nonprofit organizations use social media. *Journal of Computer-Mediated Communication*, 17(3), p. 341. doi:10.1111/j.1083-6101.2012.01576.x

9 Guo, C., Saxton, G., 2012. Tweeting social change: How social media are changing nonprofit advocacy. *Nonprofit and Voluntary Sector Quarterly*, 41, p. 61.

10 OECD, 2001. *Citizens as partners: OECD handbook on information, consultation and public participation in policy-making.* Paris: OECD Publications Service, pp. 15–16.

11 Goal 16. *Promote peaceful and inclusive societies for sustainable development, provide access to justice for all and build effective, accountable and inclusive institutions at all levels.* See more here. https://unstats.un.org/sdgs/indicators/indicators-list/

12 SDG Target 16.6. *Develop effective, accountable and transparent institutions at all levels.* See more here. https://unstats.un.org/sdgs/indicators/indicators-list/

13 Xprize. *Op Cit.*

14 The terms "civil society organizations (CSOs)" and "non-governmental organizations (NGOs)" are used interchangeably to describe the same reality. "NGO" corresponds to the name used by UN Environment.

15 UNEA, 2018. http://web.unep.org/environmentassembly/node/40734

16 UNEA, 2018. http://web.unep.org/about/majorgroups/resources/list-accredited-organizations

17 UNEA, 2018. http://web.unep.org/about/majorgroups/modalities/accreditation [Accessed between 2 and 6 July 2018].

18 UN Environment, 2018. *Accreditation criteria.* www.unenvironment.org/civil-society-engagement/accreditation

19 WWF global website. http://wwf.panda.org

20 Greenpeace global website. www.greenpeace.org/international/

21 TNC global website. www.nature.org/

22 AWF global website. www.awf.org

23 Sierra Club global website. www.sierraclub.org

24 NRDC global website. www.nrdc.org

25 Ocean Conservancy global website. https://oceanconservancy.org

26 IFAW global website. www.ifaw.org/international

27 World Animal Protection global website. www.worldanimalprotection.org

28 For the TEMA website, the English version of the website content was considered. www.tema.org.tr

29 CI global website. www.conservation.org/
30 Earth Day global website. www.earthday.org
31 EDF global website. www.edf.org
32 BirdLife global website. www.birdlife.org
33 The choice was made to focus the analysis on the global website of each organization when the organization has several regional websites. When the "global" version of the website content was not specified, the choice was made to focus on the default option, which most of the time was the USA version in English.
34 The United Nations also proposes an e-Participation index. However, it is developed to assess governments and is not based on a web content analysis, but on UN statistics. This research and this index are both based on a web content analysis.
35 This methodology is based on a web content methodology that was first applied by Pina et al. (2007, 2010) and Torres et al. (2006) in analyzing local government websites, and more recently by Acerete, Yetano, and Royo (2016) in analyzing the websites of the environment departments of local European government signatories of the Aalborg+10 Commitments. It was then adapted to the specificities of CSOs in the environmental sector. This version is specifically designed for CSOs and therefore differs from the original methodology. It consists of 65 items in total and four dimensions to focus on some specific elements of participation.
36 Acerete, B., Yetano, A., Royo, S., 2016. Evaluating public (e)information provision in Aichholzer, Georg, Kubicek, Herbert, Torres, Lourdes. ed. *Evaluating e-participation: Frameworks, practice, evidence, public administration and information technology V.19*. Cham, Switzerland: Springer International Publishing, p. 72.
37 Ibid.
38 Dingwerth, K., 2014. Global democracy and the democratic minimum: Why a procedural account alone is insufficient. *European Journal of International Relations*, 20(4), p. 1130. https://doi.org/10.1177/1354066113509116,
39 See WWF. http://wwf.panda.org/organization/finance/
40 Misuraca, G., Broster, D., Centeno, C., 2012. Digital Europe 2030: Designing scenarios for ICT in future governance and policy making. *Government Information Quarterly*, 29(1), p. 126. https://doi.org/10.1016/j.giq.2011.08.006
41 See chapter on social media for more information.
42 Sierra Club Foundation. www.sierraclubfoundation.org
43 Rosenau, T., 2012. People on the internet as agents of change. In Brousseau, E., Marzouki, M., Méadel, C. (ed.). *Governance, regulation and powers on the internet*. Cambridge, UK: Cambridge University Press, p. 117.
44 Ibid, p. 116.
45 Ibid, p. 118.
46 Latham, R., Saskia, S., 2005. *Digital formations: IT and new architectures in the global realm*. Princeton, NJ: Princeton University Press, p. 9.
47 Ibid, p. 10.
48 Padovani, C., Pavan, E., 2016. Global governance and ICT: Exploring online governance networks around gender and media. *Global Networks*, 16(3), p. 351. https://doi.org/10.1111/glob.12119
49 SDG Target 16.6. *Develop effective, accountable and transparent institutions at all levels*. See more here https://unstats.un.org/sdgs/indicators/indicators-list/
50 Xprize. *Op Cit.*

3 Social media platforms and the advocacy role of environmental civil society organizations

1. Introduction

The environment has become over recent decades a central issue in the press, but also at governmental and individual levels. Since the environment is a global issue with very concrete, tangible, and local effects on our everyday lives, every individual and organization is concerned and undoubtedly has a point of view. This is the case in Europe with the recent heated debates about pesticide reduction, and particularly glyphosate, which has led to a media buzz or virality on social media,[1] where a wide range of individuals and organizations debate and share points of view that can in turn influence public opinion and global environmental governance.

Increasingly, the problems debated and discussed in the public sphere are transnational in nature, climate change for instance. The participation of civil society has evolved with the emergence of these new challenges. The environment is an ideal topic to examine how civil society organizations (CSOs) adopt social media instruments, since it affects individuals from all parts of the world, as well as organizations at the local, national, regional, and international levels: farmers, citizen-consumers, CSOs, international networks of activists, and journalists, among others. CSOs represent a category of stakeholders that is multiple and heterogeneous, which suggests an infinite number of perspectives, approaches, and interests when it comes to the protection of the environment.

The dynamics among global environmental governance stakeholders on social media remain unclear. This is in particular due to the vast number of sub-topics, which trigger an equally vast number of communities with their own platforms, instruments, roles, rules, and dynamics. For this reason, scholars have shown an increasing interest in studying social media. An analysis of Scopus results showed that investigations with "social media" AND "politic* OR govern*" increased by 692% from 2010 to 2013, compared to an increase of 9% on the same search without "social media."[2] Nevertheless, there is still a need for integration and theoretical reflection to better understand social media as a new field of global governance.[3]

Although CSOs are often confined to observer status, their role in global environmental governance is crucial at many levels: they implement field projects and provide services, produce scientific knowledge, monitor governmental actors and multinationals, advocate for new policies, take part in international negotiations, and raise awareness about issues related to the protection of the environment. This chapter focuses on one of these roles: advocacy. Advocacy is defined as the action of publicly supporting or recommending a specific cause, policy, or idea on behalf of others.[4] The objective of advocacy is to influence decision makers and the political agenda at international, national, and local levels on behalf of other stakeholders. This is one of the main activities of civil society – for two reasons. First, CSOs are rarely granted the official status to participate in formal policy- and rule-making processes. Second, CSOs represent (or claim to represent) a group of stakeholders with specific interests. In terms of representation, social media platforms allow direct interaction with the stakeholders the organization aims to represent. Furthermore, due to the complexity and wide range of actors and interests, advocacy also plays a major role in global environmental governance. All stakeholders involved advance their interests and pursue some of their objectives through advocacy.

Providing broader access to information, consulting with the base audience, and mobilizing stakeholders correspond to the three key communicative functions for CSO advocacy work[5]: "information" (providing information relevant to one's audience), "community" (interacting with one's audience) and "action" (calling for action, also called mobilization).[6] These three functions were also identified by the OECD in 2001 as three forms of participation: "information" (access to useful and accurate information), "consultation" (possibility to co-design documents), and "active participation" (taking part in governance processes).[7] This means that they also support Sustainable Development Goal (SDG) 16, "Peace, justice and strong institutions,"[8] focusing on good governance and in particular SDG Target 16.6, "Develop effective, accountable and transparent institutions at all levels,"[9] and SDG Goal 17 entitled "Revitalize the global partnership for sustainable development."[10] Their advocacy strategies should also allow CSOs to better understand the stakeholders and the cause they represent, to gain visibility on the international stage, and therefore to contribute to the emergence of a global partnership for the protection of nature.

This chapter will first discuss social media platforms and advocacy tactics on social media, the methodology used, and then analyze how some environmental CSOs accredited to the United Nations Environment Assembly (UNEA) use their social media platforms to provide information, consult with their audiences, and mobilize them to support their advocacy strategies and promote their cause on the international stage. Hence, advocacy is one of the key roles of CSOs that is greatly affected by the generalization of social media platforms.

2. Social media platforms

If some organizations developed a website already in the 1990s, social media platforms appeared in the first decade of the 21st century. Their adoption rate by civil society was much faster than for websites for instance, since they do not represent the same amount of financial and human resources required to set up. Although today to create a website is fairly easy and affordable, it was not the case up until recently. And this is why for many newer and smaller CSOs, the choice is often made to first create a Facebook page for instance, and then at a later stage to develop a website, which can require buying a unique domain name, learning how to use a Content Management System (CMS), choosing a specific design, and producing new content.

Because they have been gradually and massively adopted by all actors, and in particular individuals, social media platforms constitute multiple online spaces, where traditional stakeholders such as states and international organizations (IOs) interact and compete with non-state actors, including individuals, civil society organizations, and businesses at the local, national, and international levels. In 2017, it was estimated that there were between 2.5 billion and 3 billion social media users around the globe.[11] A wide array of environmental actors use social media for advocacy purposes, enabling them to set the international agenda, to organize group actions and collaborate on a global scale, and finally to collect information for advocacy purposes, whether marketing or public relations.

Social media platforms can be characterized by their ability to enable many-to-many communication on a global scale. They can be described as a collection of instruments. Previously, interpersonal communication was possible only on dedicated media such as telephones. Similarly, traditional broadcasting media such as TV, radio, and newspapers delivered information to a large number of people, but this was one-way communication only. Social media is an innovation first in the sense that it blends one-to-one with one-to-many communication streams, enabling individuals to become producers of information and not only consumers, and allowing broadcast media to target their messages individually. Second, social media is an innovation because it allows two-way communication, where organizations and individuals interact on a permanent basis on social networking platforms such as Facebook, Twitter, and LinkedIn. Relationships are at the core of social media platforms.[12]

Among those who benefited most from the generalization of ICTs and more precisely from computerization, internet penetration, and the rapid adoption of mobile phones are the individuals and CSOs that gained new capacity to reach out to a larger range of stakeholders,[13] such as donors, volunteers, online and offline media, and the general public. The diversity of the environmental community provides an excellent opportunity to examine the use of social media, which differs from one organization to another. Actors of global environmental governance have developed a capacity for outreach in order to gain new donors, members, and volunteers. In a knowledge society, information becomes both a resource and a force that transforms decisional and non-decisional processes. Many CSOs have developed web 2.0 public relations campaigns to raise awareness about specific issues, which has triggered the emergence of new online spaces where legitimacy is built through the inclusion of a wider number of stakeholders, starting with the general public. Citizens become actors on the international stage by interacting with CSOs, IOs, states, and businesses on social media. In that context, social media platforms can change how global environmental governance actors relate to each other, and the dynamics of their interaction.

This gradual technological shift has also dramatically transformed the media landscape. To name only a few, traditional media, such as printed matter, had to reinvent themselves, adding an online component to the paper version. Most newspapers, if not all of them, have a website, social media channels, and specific content developed for their online audiences. Some citizens have become journalists and started reporting on events from the field. Social media, and in particular Twitter, has become the place to be when it comes to finding or publishing the latest information. The most recent news items are no longer found in newspapers but online. In a large number of countries, citizens read news predominantly on social media.

Web 2.0 is based on the new capacity offered to users to self-generate content online and to interact with other users without the interference of elite media and traditional sources of authority. The previous version of the web allowed only one-way communication; the online user was a passive consumer of information. With web 2.0, online users have become consumers and producers of information.

This rapid transformation of the media landscape became of high interest for researchers and policy makers, since this change affects the role and power of media. In particular, a considerable number of investigations analyze the use of social media by CSOs.[14] and most argue that CSOs still need to develop additional capacity in order to gain the most out of social media platforms.[15] It seems indeed that CSOs have not fully embraced the new two-way communication paradigm that a presence on social media implies.[16] This is one of the elements this chapter aims to discuss.

Social media platforms can be categorized according to their geographical scope and utility. On the one hand, "universal" social media platforms, such as Facebook, provide users with a new digital space on a global stage where they can create either personal profiles or an official page for an organization. This type of platform allows for personal but also for professional interactions. Intimate conversations can take place, as well as business transactions. Users write comments, share views, post and watch videos, play games, respond to quizzes, etc. On the other hand, some specialized social media platforms cater to either dedicated geographical parts of the world population (e.g. WeChat in China and some parts of Asia) or some professions (e.g. LinkedIn for professionals or ResearchGate and Academia for academics), while others offer tools for specific activities, such as publishing videos (e.g. YouTube), photos (e.g. Instagram), or text content (e.g. Blogger), among others. Of the social media platforms, Twitter is a microblogging application where individuals and organizations publish tweets – short messages.[17]

Through its multiple platforms, social media allows personal interactions, peer recognition, and the strengthening of group norms[18], which in turn stimulate individual and community identity construction, two crucial components of political conduct. The various aspects linked to the environment, such as use of land, food safety, and ecosystem management, among others, are crucial components of identity construction, and have led to the emergence of numerous online communities. For instance, the grassroots movement "Right To Know Rally" began its activities with one page on Facebook[19] and rapidly grew into an international movement spread over 400 cities in North America and Europe. Furthermore, individual forms of online protests are increasingly associated with lifestyle elements, which results in the personalization of global issues.[20] This implies that in the war of narratives on social media platforms between multinational corporations, governments, and the global civil society, the framing of narratives becomes central and influences the definition of global public issues.

3. Social media and advocacy

According to the Oxford online dictionary, advocacy means to publicly support or recommend a particular cause or policy.[21] Three key communicative functions are used by nonprofits for their advocacy work on social media. First, *information*[22] is when the organization presents itself, its activities, and future events and provides information that is relevant to its audience. Second, *community*[23] is when the organization interacts with its audience and aims to develop a community. Lastly, *action*[24] is when the organization sends out a call for action such as to participate in an event, donate, or share a type of media. This is not the only way to categorize social media tactics, but one that corresponds to a general understanding of the three main types of actions found on social media.

Another approach similarly divides these tactics into three similar categories, but with different names. First, *reaching out to people*[25] corresponds to the information function in the previous model. Advocacy work aims to educate and raise awareness about specific issues. Second, *keeping the flame alive*[26] corresponds to community and dialogue. It allows organizations to deepen existing relations with their audiences and develop new ties with others. The organization aims here to nurture its audience and build an active community of supporters. Third, *stepping up to action*[27] corresponds to the organization calling for action and asking its supporters to mobilize. What is interesting to see is the convergence between the different analyses performed on social media, which indicates that these three categories effectively represent how CSOs manage their social media platforms.

CSOs have increasingly adopted technologically intensive media to influence various stakeholders;[28] among the most commonly used are social media platforms. Indeed, individuals are more easily approached through these new instruments for a number of reasons. First, social media users can see at the same time and in the same place a mix of personal contact updates and sponsored content that promotes a product, a service, or a CSO. This means that the line between personal and intimate content and sponsored content is blurred, which can greatly benefit advocacy campaigns in terms of credibility and trust. One naturally has more trust in information distributed in a secure and intimate space than in the outside world.

Second, individuals regularly visit social media channels. According to Pew Research Center, Facebook usage and engagement were still on the rise in 2016, while the adoption of other platforms held steady. On a total population basis (taking into consideration non-internet users), 68% of all US adults use Facebook, 28% Instagram, 26% Pinterest, 25% LinkedIn, and 21% Twitter. Moreover, in terms of the frequency of use of social media platforms, 76% of Facebook users, 42% of Twitter users, 51% of Instagram users, 25% of Pinterest users, and 18% of LinkedIn users visit the site at least once a day.[29] In other words, the vast majority of the population connects daily to social media platforms, which become a one stop shop for CSOs and other organizations to promote their content and services.

Third, social media offers a large number of promotional instruments, ranging from ads with highly detailed and targeted segmentation tools, to contests and customized applications. Social media allows CSOs to rapidly and efficiently identify their target audiences, organizations with common agenda, and empathetic individuals. For advocacy professionals, this is a gold mine, since they can segment their audiences at a level unforeseen previously. Fourth, each element of a campaign, whether an application, a message, a post, or an ad, can be designed for a specific part of the population, and can further be monitored and readjusted depending on the success of the campaign.

The wide variety of social media channels facilitates the exchange of user-generated text, audio, and video files and instruments, and has empowered individuals and organizations to develop and conduct advocacy campaigns.[30] Social media platforms rely mostly on the content generated by organizations and individual users,[31] where organizations and individuals meet online, and where CSOs strive to attract new stakeholders such as potential donors, members, volunteers, petition signatories, and digital ambassadors who can convey their message to their own personal networks. This leads to the creation of online like-minded communities with common interests and agendas that can support the work of CSOs and spread their messages further afield.

When an event is communicated by an organization, the next stage is to relate this event to other events; this then becomes a news item and creates momentum, a buzz, that provides a space for many actors to intervene and position themselves. Their responses become news, leading to other comments and responses. This self-reinforcing process is strengthened by the networking nature of social media platforms. Indeed, citizens are more and more often the ones recording videos and taking photos that are posted on social media and that feed the momentum.[32] User-generated content greatly increases the availability of information. Also, group formation and the personalized framing of information that are specific to social media make these platforms echo chambers where an issue can escalate and become viral.

Moreover, emotional messages and images attract the attention of users on social media,[33] and can increase the momentum for a specific issue. This is particularly true when health is threatened by the issue at stake, illustrated by tragic images of individuals suffering. A news item will tend to spread across numerous online clusters and networks, with comments added to original framing. In certain cases, however, some discourses are restricted to specific communities, such as extreme left or extreme right political messages.

The liking, commenting, and sharing on social media trigger a rapid dissemination of a news item. Social media channels offer a platform for citizens, consumers, businesses, politicians, and experts to have a say and make their voices heard. In most cases, important leaders from the private and public sectors have an account on Twitter or Facebook, which leaves them no chance to avoid participating in debates that concern them. Crucial actors and influencers are often publicly addressed on social media, and the strategy to address these requests is often complicated and personal. Crisis communication is a science and an art, especially on social media. A good answer or no answer can sometimes end the controversy, whereas no answer or a bad answer can at other times trigger even more controversy. Users on social media spread information by reacting to a news item, thus contributing to the emergence of momentum. The feedback loops between a news item and the responses of social media users lead to an exaggeration of reality, often associated with media hype[34] and virality. These momenta represent a new opportunity for various social actors to influence the general public and policy makers. The dynamics associated with the global character of the environment and social media platforms imply that local news can quickly become a global issue.

4. Methodology

Since this chapter examines the use of social media by CSOs in global environmental governance, the choice was made to focus on the non-governmental organizations (NGOs)[35] that UN Environment has accredited to the United Nations Environment Assembly (UNEA). UNEA was created at the 2012 United Nations Conference on Sustainable Development, often referred to as the Rio+20 Conference. UNEA's Members adopt resolutions and initiate calls to action in order to coordinate intergovernmental initiatives related to the environment. They meet every two years to address global environmental challenges, set priorities for new environmental policies, and adopt international environmental law.[36]

Although UNEA Member state representatives are the only ones allowed to vote, major non-governmental groups and stakeholders are granted access and observer status to UNEA. More precisely, the Assembly has accredited organizations from the following groups:[37] NGOs (322), the scientific and technological community (34), business and industry (26), children and youth (25), women (20), indigenous people and their communities (13), farmers (5), local authorities (4), and workers and trade unions (4).[38] Observer status allows accredited organizations to participate as observers in the Plenary, the Committee of the Whole, and the Ministerial Consultation's discussions. They can also circulate written statements to state representatives through the UN Environment Secretariat. Finally, they can make oral statements during the discussions of the UN Environment Assembly.[39] To be accredited to the Assembly, non-governmental organizations must show proof that they adhere to the following criteria: engagement in the field of the environment, international scope of work, nonprofit, officially registered as a legal entity in at least one country, and at least two years of existence.[40]

Since UNEA is the highest-level political body of global environmental governance, organizations accredited to the Assembly represent the right sample of non-governmental organizations susceptible to using social media to advocate and fundraise. Indeed, these accredited organizations undoubtedly aim to influence the international agenda and global environmental governance decision makers. Therefore advocacy is one of their main objectives: they would not make the effort to become accredited otherwise. Their use of social media should reflect this objective. Also, by joining UNEA as observers, these organizations seek a global audience or at least global exposure, since they will be interacting with states and international organizations. Hence, they are a sample of international civil society organizations that aim to gain visibility and influence on the international stage. This chapter examines their use of two social media platforms to reach these objectives.

This chapter aims to provide an overview of the main advocacy tactics used by CSOs on social media. Indeed, as described previously, CSOs use social media to increase their influence on the international stage and set the agenda. They do so by raising awareness about specific environmental issues, developing a community and interacting with it, and finally mobilizing it to put pressure on governments and decision makers. At UNEA, CSOs that become accredited must show proof of the international scope of their work. This means that their audiences and/or their topics of interest must be global. Their choice to become accredited to UNEA reflects their interest in influencing the international stage and in particular states. Therefore they must have developed strong advocacy competences and tactics; they should in fact represent some of the best examples in global environmental governance in terms of advocacy tactics on social media since they chose to join UNEA with an observer status.

Two analyses were conducted to examine the social media strategies or tactics of NGOs accredited to UNEA. First, an analysis of the presence of the NGOs accredited to UNEA on social media examined how many organizations adopted and used Facebook and Twitter. This included examining how many organizations are present on Facebook and Twitter, how they communicate on social media, the number of followers and likes, and how they present themselves on these platforms. The choice of Twitter and Facebook is due to the fact that these two platforms are quasi universal in terms of geographical scope and they are the platforms most frequently used by the NGOs accredited to UNEA. YouTube, Instagram, Google+, and LinkedIn are also present, but in a much lower proportion. The presence of the 322 NGOs on Facebook and Twitter was examined over a one-day period in July 2018. The objective of this first analysis was to provide an overview of the presence of accredited NGOs and inform or confirm the dominant narrative that hints that social media instruments have become an integral part of the advocacy strategies of CSOs, due to their low cost of entry.

Second, a content analysis of CSOs' activities on Facebook and Twitter during the month of June 2018 was performed. It focused on the way accredited NGOs communicate on social media: what type of media is used, what content is produced, how their audiences react to these communications. This should provide a good indication of how these types of organizations use social media since they are the most active ones. The choice of timeframe corresponds to prior social media investigations.[41] The month of June was selected because it is not a holiday period and therefore it should illustrate "business as usual" activities. Furthermore, this month encompasses several nature-related world days, including World Environment Day, as illustrated in the following examples. Hence more online activities can be expected around this date.

Since this second analysis required analyzing manually a large number of publications, a smaller sample of NGOs was chosen. The choice was based on the language of publication (English), their main focus of work (protection of the environment) and the number of likes and followers on Facebook and Twitter platforms. Some of the organizations accredited to UNEA also include humanitarian organizations such as the International Committee of the Red Cross (ICRC). Since this chapter aims to illustrate how environmental organizations use social media platforms, they were not included in this small sample. Choosing organizations with the largest audience implies that these organizations can be considered "champions" of advocacy on social media, which should illustrate best how social media platforms can be used to advocate. Among the organizations with the largest audience, five organizations were chosen to illustrate different contexts: a regional scope (Africa Wildlife Foundation), a specific environmental issue (Ocean Conservancy), an activist perspective (Greenpeace), and a more "generalistic" approach (WWF international and the Nature Conservancy). This categorization of the five organizations chosen for this sample does not represent the complexity and wide scope of their actions and roles on the international stage. It is rather to explain what motivated their selection: the intention to offer some diversity and illustrate different strategies and approaches to advocacy among the organizations with the largest audience on social media platforms.

A coding was applied to all publications on Facebook and Twitter in June 2018. They were sorted into the three categories of advocacy tactics described previously: information, community, and action. Although coding can be subjective, two rounds of coding taking place at two different times were conducted to categorize the content published between 1 and 30 June 2018. The main objective was to identify and illustrate some advocacy strategies and understand how five of the environmental NGOs most visible on Facebook and Twitter communicate with and advocate to their audiences.

As discussed previously, social media platforms offer CSOs affordable tools to reach out to their audiences and therefore gain new skills to raise funds and advocate. In terms of social media use, Facebook and Twitter are the most represented and used among all 322 organizations, along with LinkedIn, YouTube, and Instagram. To follow is the analysis of social media presence and advocacy tactics on Facebook and then on Twitter.

5. Social media presence

The analysis of the presence of the NGOs accredited to UNEA focused on two social media platforms: Facebook and Twitter. Facebook distinguishes between an individual user's profile and an organization's page. Although some organizations have several pages for their regional offices or for specific activities, this analysis focuses on their main official page. When an individual user likes the official page of an organization, he/she will receive the latest updates and publications from this organization in his/her newsfeed. However, the newsfeed is managed by an algorithm that decides what information is shown in which priority in each newsfeed; this is based not only on what the organization publishes, but also on a number of criteria such as the number of likes the publication has received, the level of closeness with the organization, what other profiles and pages publish at the same time, etc. This means that it might be complicated to ascertain precisely who sees what on Facebook.

Out of the 322 NGOs, 69% have a Facebook page. Facebook is a platform that allows direct interaction with stakeholders and in particular with the general public. It enables organizations to promote their work and to raise awareness about specific issues directly with individuals. This accessibility can also attract non-institutional donors. Facebook is an ideal platform to showcase the work of an organization and to use multimedia such as video, photos, and text to develop a communication campaign at a low cost. Facebook allows these organizations to promote their work and target a very specific segment of their audience based on demographics such as age, gender, location, but also tastes and hobbies. One can promote a publication or page to individuals who like a similar page. This means that Facebook will be used mainly to connect with others either to inform or to call for an action, which could include such things as signing a petition or participating in a protest.

In terms of geographical distribution, West Asia and Latin America and the Caribbean have the strongest presence, with respectively 82% and 86% of their organizations having a Facebook page. Europe and North America follow with 70% each, and lastly Africa and Asia Pacific, both with 60%. The lower presence of Asia Pacific NGOs on Facebook can be explained by the fact that this region has developed its own set of social media platforms such as WeChat, Weibo, QQ, and Baidu, to name only a few, meaning that many of these organizations have a presence on social media, but not on Facebook. On the other hand, Africa has not developed its own social media platforms, and therefore the low levels of social media presence merely reflect the digital divide.

In terms of identity, the top 15 environmental NGOs accredited to UNEA and with the most users liking their Facebook page[42] all have a well-documented section "about" the organization on their Facebook page. They represent the organizations with more than 300,000 users who like their official page. However, the difference in terms of audience is substantial between the first and the fifteenth organization of this ranking: WWF has an audience ten times larger than BirdLife International's. This illustrates the considerable difference between levels of presence on social media. Among the top 15 are nine from North America, five from Europe, and one from Africa, which illustrates once again the digital divide between different parts of the world.

Among the top 15 environmental NGOs, the large majority clearly mention the term "advocacy" or synonyms such as "influence," "empower," or "campaigning." Examples include "campaign on a variety of issues (. . .),"[43] "(. . .) influential grassroots environmental organization,"[44] "(. . .) effective environmental action group, combining the netroots power of more than 2 million members and online activists,"[45] and "(. . .) by campaigning against the commercial hunt for whales and seals."[46] The other organizations present their mission in more general terms without an explicit mention of any term directly related to advocacy. WWF is a good example where advocacy is not mentioned in its mission, but is implied with an injunction to act in a certain manner: "We need to stop damaging our only life support system. (. . .) We need to share our world with all the other species that live on it. (. . .) And we must stop being the cause of their disappearance. (. . .) We must learn to live with what natural resources are available to us."[47]

This is not to say that the other organizations do not advocate, but rather they do not describe their mission in these terms. They also use action verbs to describe their activities such as "to stop the degradation of the planet's natural environment and to build a future,"[48] "to conserve the lands and waters on which all life depends,"[49] "to ensure wildlife and wild lands thrive in modern Africa,"[50] "protect the ocean from today's greatest global challenges."[51] In other words, the majority of the top 15 environmental NGOs accredited to UNEA with the most likes on Facebook identify one of their objectives as being advocacy.

In terms of audience, these 15 environmental organizations have accumulated a substantial number of users who like their official page ranging from 3,071,977 for the World Wide Fund for Nature (WWF) to 320,861 for BirdLife International, as shown in Table 3.1. With 15,834,283 "likes," the top 15 environmental organizations represent more than 74% of the total of 21,219,612 "likes" of the official pages of the 322 NGOs accredited to UNEA. The top three organizations (the World Wide Fund for Nature, Greenpeace International, and The Nature Conservancy) account for more than 34% of the total 21,219,612 "likes" for all NGOs accredited to UNEA and 46% of the top 15 environmental NGOs. This shows how social media presence is in fact quite concentrated in the hands of a small number of major environmental organizations, which have developed a large audience, and which concentrate the capacity to attract the attention of Facebook users. This implies that they have the strongest influence on social media platforms, and therefore are the best suited to frame environmental issues and discourses, and to raise awareness based on their views and values. Hence, social media platforms have not only allowed a large number of organizations to reach out to global audiences, but also contributed to a concentration of the capacity to capture the attention of citizens, and therefore slightly reduced the plurality of opinions.

Contrary to Facebook, Twitter offers only one type of presence on its platform: there is no distinction between an organization and a user. This means that an organization can easily follow (and be followed by) a broad range of organizations

Table 3.1 Top 15 environmental NGOs accredited to UNEA with most Facebook likes

Region	Name of accredited NGO	Number of likes[52]
Europe	World Wide Fund for Nature (WWF)	3,071,977
Europe	Greenpeace International	2,917,370
North America	The Nature Conservancy (TNC)	1,354,078
Africa	African Wildlife Foundation (AWF)	1,284,310
North America	Sierra Club	998,119
North America	Natural Resources Defense Council (NRDC)	929,798
North America	Ocean Conservancy	890,541
North America	International Fund for Animal Welfare (IFAW)	715,228
North America	Defenders of Wildlife (DOW)	687,507
Europe	World Animal Protection	620,932
Europe	The Turkish Foundation for Combating Soil Erosion for Reforestation and the Protection of Natural Habitats (TEMA)	580,477
North America	Conservation International Foundation (CI)	375,409
North America	Earth Day Network (EDN)	374,084
North America	Environmental Defense Fund (EDF)	338,177
Europe	Birdlife International	320,861

and individual users. Each profile has a Twitter handle, @WWF for example, which corresponds to the name of the user of the platform. Some organizations have in fact several accounts and Twitter handles for their regional offices or for specific activities. This analysis focuses on their main official account and handle.

Twitter is used not only to be found, but also to send direct (hidden from other users) messages to a person and to mention this person in a tweet. Furthermore, Twitter makes the level of interaction between one user and another user very visible. At the top of each profile, Twitter indicates the number of users the profile follows (tab "following"), the number of users who follow the profile, and the number of tweets the profile likes (tab "likes"). This provides a good overview of how interactive the user is with his/her audience (likes) and with other users (following). In addition, Twitter allows the creation of # hashtags that reference content according to topics (e.g. #biodiversity). A Twitter list is a curated group of Twitter accounts that shows a stream of tweets from only the accounts on that list.[53] In some countries, Twitter also allows users to create "moments," which are curated stories to showcase a specific event.[54]

For Twitter, the distribution differs slightly from Facebook results: 59% of all accredited NGOs have an official account on Twitter. Twitter has become the foremost source of news for millions of users. It connects people throughout the world and allows them to share the latest information on many topics. Trends and influencers are closely followed by many journalists in order to report the latest piece of information. Although Twitter also connects individuals, its use differs substantially from Facebook. In very simplistic terms, Facebook could be more associated with leisure and entertainment, whereas Twitter has a more "professional flair" as it is also largely used to distribute formal information. For many

NGOs, Twitter can become a very useful tool to advocate for specific causes and influence the debate, reach out to influencers, and become better acquainted with the arguments of all parties to a debate.

This implies however that the use of Twitter requires additional skills and time dedicated to identifying the right influencers, hashtags, and accounts, and connecting with sources of information to try to have content shared by influencers. Indeed, as on Facebook, it is not possible to promote content through paid posts and ads. Furthermore, the results of these actions might not be as visible as on Facebook since part of the use of Twitter is to gather information and influence others, contrary to Facebook, which is more conducive to community building and fundraising activities.

In terms of geographical distribution, Europe (74%) and West Asia (71%) are the regions with the highest presence on this platform, followed by North America (62%), Latin America and the Caribbean (57%), Africa (52%), and Asia Pacific (35%). Here as well, the fact that Asia Pacific has developed its own social media platforms explains the low presence of their NGOs on Twitter.

In terms of identity, the top 15 environmental organizations accredited to UNEA with the most Twitter followers[55] have a short description of their activities and a link to their websites. However, the descriptions are much shorter than on Facebook, due to the specificity of this platform.

In terms of audience, the numbers of followers of these top 15 environmental organizations range from 4,030,000 (WWF) to 56,400 (EDN) as shown in Table 3.2. The top 15 environmental NGOs accredited to UNEA with the most Twitter followers account for 68% of the total of followers of the 322 NGOs accredited to UNEA. These 15 organizations are the same (but in a different order) as the top 15 environmental organizations with the most users liking their official Facebook page. In other words, the geographical distribution is similar, with only one organization from Africa. This shows once more the digital divide, and is concerning in terms of plurality of sources of information on social media platforms.

Moreover, the top three organizations with the most followers on Twitter are the same organizations with the most users liking their official Facebook page, and in the same order: the World Wide Fund for Nature, Greenpeace International, and The Nature Conservancy. They account for more than 50% of all followers of the 322 NGOs and 75% of the top 15 environmental NGOs. These results confirm that social media presence is quite concentrated in the hands of a small number of major environmental organizations, which have developed a large audience, and which concentrate the capacity to capture the attention of Twitter users. As mentioned previously, these organizations have the strongest influence on social media platforms, and therefore are the best suited to frame environmental issues and discourses and raise awareness based on their views and values.

Although one user can follow many organizations at the same time, this demonstrates well that these organizations have acquired most of the presence on Twitter and concentrate most of the attention on this platform. The information they provide will reach out to a very large audience, and therefore they represent the main sources of information and reference for users interested in nature conservation.

Table 3.2 Top 15 environmental NGOs accredited to UNEA with most Twitter followers

Region	Name of accredited NGO	Number of followers[56]
Europe	World Wide Fund for Nature (WWF)	4,030,000
Europe	Greenpeace International	1,740,000
North America	The Nature Conservancy (TNC)	925,000
Europe	The Turkish Foundation for Combating Soil Erosion for Reforestation and the Protection of Natural Habitats (TEMA)	409,000
North America	Sierra Club	329,000
North America	Ocean Conservancy	314,000
North America	Natural Resources Defense Council (NRDC)	297,000
North America	Environmental Defense Fund (EDF)	175,000
North America	Defenders of Wildlife (DOW)	145,000
North America	Conservation International Foundation (CI)	143,000
Europe	World Animal Protection	121,000
Africa	African Wildlife Foundation (AWF)	112,000
North America	International Fund for Animal Welfare (IFAW)	74,600
Europe	Birdlife International	66,700
North America	Earth Day Network (EDN)	56,400

6. Social media content analysis

As mentioned previously, the three top NGOs accredited to UNEA and that have the largest audiences on Facebook and Twitter are the World Wide Fund for Nature, Greenpeace International, and The Nature Conservancy. As they represent the organizations with the strongest presence on both social media platforms, they were chosen to analyze in more detail their advocacy tactics. The World Wide Fund for Nature and The Nature Conservancy seem to target a more general audience, whereas Greenpeace has a more activist and environmental audience. In addition to these three organizations, two were added from the top 15 organizations with the biggest audience on Facebook and Twitter. The objective was to add an organization with a regional scope, and an organization with a topical scope: African Wildlife Foundation has a regional scope, and Ocean Conservancy focuses on oceans. Both organizations tweet in English, as the three previous ones.[57] In terms of advocacy tactics, they have adopted distinctive strategies with specific objectives, tone of voice, and story line. One common element is their wide use of images and videos. Almost all posts and tweets contain either one or the other.[58]

With a total of 43 posts over a one-month period, WWF is not the most active organization on Facebook, although it has the most likes on its official page. The story line is the beauty of nature and the urgent need to protect it. Its emblem is quite representative of these two points: the beauty and yet vulnerability of the Panda. It raises awareness about the necessity to protect the environment by focusing on a positive view of the environment and on what it brings to human beings: ecosystem services such as providing food or oxygen. Hence, its discourse frames nature as a treasure for citizens to acknowledge and protect. It does not pursue "name and shame" communication and does not address directly any government or multinational corporations that might have committed some environmental damages. It rather focuses on what citizens can do to better value nature and protect it.

The organization focused primarily on two tactics in June 2018: information (40%) and action (37%), indicating that the organization uses Facebook to inform and mobilize its audience. This is quite interesting since previous findings[59] indicated that the action tactic was the one least developed on social media. This point also reflects the fact that WWF does not need to build its community since it is already vast and interested in the topic. In June, WWF mobilized its audience toward reducing consumption of plastic and helping preserve the oceans from plastic waste. For instance, on World Oceans Day (8 June), WWF wrote: "From providing food to oxygen, the ocean does so much for us. But plastic pollution is threatening the health of our oceans and marine life :(Land to sea, we're all connected. This #WorldOceansDay, take action to #UseLessPlastic: https://pand.as/WED-2018 #connect2earth."[60] Also relating to plastics, the most shared Facebook post was published on 5 June: "Do you need more reasons to #UseLessPlastic? Show your support for our shared home this #WorldEnvironmentDay! From the air we breathe to the water we drink, nature does so much for us. Here's how you can do your part: https://pand.as/WED-2018 #BeatPlasticPollution #Connect2Earth"[61] and included a video about plastic waste and its impact on the ocean. Posts with the most "likes" or "shares" are either informational or calls for action. The tone of voice is fairly friendly. Very few posts trigger the use by the audience of the "laughing" emoji. The ones most used are "like," "love," "sad," and "angry," which corresponds well to WWF's story line.

On Twitter, WWF advocacy has been developed with a similar strategy, tone of voice, and story line. With a total of 42 of its own tweets and the largest audience, WWF focuses its communication mainly on providing information to its audience (62%) such as on 5 June: "HISTORIC: Mexican President @EPN announces protection of nearly 300 river basins, guaranteeing water supply for the next 50 years to 45 million people and some of the most biodiverse ecosystems in the country 🎉😊🐋#WorldEnvironmentDay https://pand.as/2xIvuVF."[62] The topics addressed inform users about the protection of species and ecosystems using beautiful photos and videos. In terms of action, WWF focused on using less plastic and mobilized its audience to use less plastic (e.g. on 5 June: "Do you need more reasons to #UseLessPlastic? Show your support for our shared home this #WorldEnvironment! #connect2earth & do your part for 🌏: https://pand.as/WED-2018 #beatplasticpollution").[63] In terms of community, WWF engages with its audience by using hashtags and stories to tell (e.g. "Today marks 100 days to the Global Climate Action Summit, and we're celebrating by highlighting inspiring climate solutions. Share your favourite solutions from around the world with #GCAS2018! #StepUp2018").[64] The tweet with the most likes (2000) was published on 22 June and showcased an endangered species: "Exciting news! A young male #leopard has found his way into #Armenia. The last leopards in this area were sighted 18 years ago but were #poached & forced out of the region. This is truly a happy sight with less than 1,300 Caucasian leopards left in the world."[65] In terms of tactics, WWF shows that its audience is vast and probably more "general public" than Greenpeace's for instance. Therefore its advocacy tactics differ, with an editorial line focusing more on protecting the beauty of nature than on denouncing the abuses (as Greenpeace would). Needless to say, both approaches are highly necessary and complementary.

The second environmental NGO accredited to UNEA with the strongest presence on Facebook and Twitter platforms is Greenpeace International. With 117 posts on Facebook over a period of one month, this organization is by far the most active of this sample, with about four posts per day on average. The story line of the organization is different from that of other organizations: it focuses more on the threats against the environment and on the need for civil society to take action and defend against the environmental damages perpetrated by some governments and multinational corporations. The tone of voice is therefore quite direct, and of engaged activists, and stresses the urgency of the situation. "Name and shame" is often used. This communication strategy reflects well the history of this organization and how it was founded. Greenpeace International has positioned itself as an activist organization: images and videos of their boats intending to stop whale fishing and nuclear testing are common knowledge. Other protests against the oil and gas industry are also quite well-known. This organization has become a very vocal voice on the international stage to denounce and take action against environmental damages.

Its communication strongly reflects the lobbying and activist spirit of the organization. Here as well, the majority of advocacy tactics focuses on information (39%) and action (47%). However, contrary to WWF, where the calls for action in all cases but one incite the audience to change and adopt a more sustainable behavior, Greenpeace uses social media to mobilize its audience for a number of causes. The organization aims to promote more sustainable behavior (e.g. eat less meat on 11 June: "Celebrate #WorldMeatFreeWeek starting today! Join the movement of people choosing LESS meat and MORE thriving forests, clean air, clean water, healthy bodies & total yumminess >> https://act.gp/2KTPSWd 🍃 #lessismore"), but also to reach out to businesses and political leaders in order to influence their decisions (e.g. on 23 June: "This may look like scenes from a dystopian movie, but they are actually tar sands: a deposit of one of the dirtiest oils in the world. Join the wave of resistance to stop the expansion of tar sands pipelines: https://act.gp/2MQe2C7" or on 21 June: "the oceans are drowning in plastic waste. But thanks to you, people around the world are challenging corporations to take responsibility. Here's how you can help too >>> https://act.gp/2K7wE1W"). These examples reflect the more "activist" advocacy approach of Greenpeace. The post with the most shares and likes was on 14 June: "Happy #PrideMonth, from Mother Nature to you." It showed a video of a waterfall with the sun shining over it, producing a rainbow. It was shared more than 6,000 times and liked more than 15,000 times (all emojis of appreciation included).

With 291 tweets over a one-month period, Greenpeace is also the most active organization of this sample on Twitter. The tone of voice is more direct and specifically addresses multinational corporations, as well as political and business leaders (e.g. on 6 June: "Poisonous glop. Death-funnels. The world's dirtiest fossil fuels. Whatever you call them, tar sands oil pipelines are bad news, and @JustinTrudeau should know better www.esquire.com/news-politics/politics/a21074424/trudeau-tar-sands-climate-change-bitumen/ . . . #StopKM #Crudeau").[66] The majority of tweets are informational (more than 60%), with content about fossil fuels, plastics, waste, species and ecosystems, and climate change solutions. The organization also more directly addresses some specific issues such as the construction of tar sands oil pipelines. About 25% of its tweets focus on action. This means that it mobilizes its audience quite a lot on issues such as adopting a more vegan diet, signing petitions, and taking part in protests (e.g. on 21 June: "Our planet doesn't need more climate-wrecking oil pipelines, but banks keep financing them. Stop the money, stop the pipelines >> https://act.gp/2lnP54h #StopPipelines").[67] The tweet with the most likes was published on 8 June with a quote from Arthur C. Clarke: "How inappropriate to call this planet Earth when it is quite clearly Ocean."[68] From the tone of voice and the topics addressed, Greenpeace is talking to an audience who is already convinced of the need to act to protect the planet. Hence there is no need to focus on the beauty of nature, for instance as WWF does. The role of the organization is to support and represent civil society in the fight against political and business leaders who damage nature. There is less need for community tactics.

The third organization with the largest audience on Facebook and Twitter is The Nature Conservancy.

Its social media strategy is quite similar to WWF's, as is its audience. The story line, the tone of voice, and the audience are also about showcasing the beauty of nature with the objective to create a stronger bond between citizens and the environment. TNC raises awareness about the protection of the environment also by showing how valuable nature is for all human beings. The organization does not often mention a particular government's policies and multinational corporations. It does not focus on specific environmental damages. It is rather about reconnecting citizens with nature, educating them, and showing all the services provided by the environment.

On Facebook, the organization has more than 1,354,078 users liking its official page, and published 52 posts during the month. This is more than WWF but far less than Greenpeace International. In terms of advocacy tactics, its major focus is on information: reaching out to people. Indeed, 71% of all posts were informational, with reports about nature conservation, facts and figures, the work of the organization, sharing content from other platforms such as blogs (e.g. on 20 June: "Did you know that 2 out of 3 people will live in a city by 2050? Putting nature in cities makes cities better" from blog.nature.org),[69] and showcasing the beauty of nature (e.g. on 19 June: "The 2018 Photo Contest hinged on the curiosity of creative eyes from 135 countries. All we can say is wow").[70] With regards to community, keeping the flame alive represents 15 out of 52 posts, with several posts about photo contests showing beautiful aspects of nature. The only two calls for action ask the audience to vote in the photo contests, and therefore can arguably be considered "community tactics" more than calls for action (e.g. on June 15: "Hey you! Help us choose the 'People's Choice' winner for the 2018 Photo Contest. You can vote for as many photos as you'd like. Voting ends in a few hours!").[71] This highlights how different this social media strategy is, even though The Nature Conservancy has a similar number of likes to Greenpeace International. The post with the most shares was a post from blog.nature.org on 8 June: "'The ocean stirs the heart, inspires the imagination and brings eternal joy to the soul.' -Robert Wyland Thank you, oceans. Thank you, coral reefs. #WorldOceansDay." This is the organization that shares the most content from other sources and platforms.

With 172 tweets over a one-month period, The Nature Conservancy is much more prolific on Twitter than on Facebook. The organization tweets regularly about the same issues and sometimes with the same text and image. The general tone of voice is very different from Greenpeace International's for instance: less direct and rarely mentioning current political leaders. Rather, The Nature Conservancy focuses on specific issues such as protecting coral reefs and marine life, nature in an urban environment, indigenous people, and nature conservation techniques and advances (e.g. on 14 June: "Coral reefs aren't just pretty to look at. They're worth at least $4 billion in flood protection per year. Thank you, corals! https://usat.ly/2sR5YbB (@USATODAY)").[72] Advocacy tactics are first about providing information and raising awareness (more than 70% of all tweets), and second about community building (25%). Concerning the latter, a photo contest invited the audience to submit photo shots and participate in the voting (e.g. on 14 June: "You contributed your most meaningful nature photos. Now help us judge them! Pick your favorites for our 'People's Choice' winner in the 2018 Photo Contest: http://bit.ly/2Mo9hzw").[73] Again in terms of community building, a fun tone of voice was adopted on Father's Day with a play on words (e.g. on 17 June: "I tried to catch some fog, but I mist. 😔 #FathersDay 'Dad jokes' courtesy of @ Conserve_WA: http://bit.ly/2JMpNaF").[74] In terms of action, The Nature Conservancy mobilized its audience only in a few cases (e.g. on 7 June: "More #citizenscience opportunities for those who want to help save the rainforest from their laptops: You can easily help identify rainforest flowers without any prior experience: http://bit.ly/2sj5kU3").[75] The tweet with the most likes was published on 6 June and accurately reflects the friendly and "community" oriented tone of voice of The Nature Conservancy: "Keep calm and carry on. Smile through adversity. The glass is half full. Whatever the mantra, coral reefs need our optimism and action to save them. It's not too late. http://bit.ly/2JggYFM #WorldOceansDay #CoralOn."[76]

Similar to The Nature Conservancy, Africa Wildlife Foundation's (AWF) main advocacy tactic on social media is to provide information. The story line of this regional organization is to showcase the beauty of nature and the need to protect it. However, contrary to the four other organizations discussed here, its tone of voice is more formal. Its communication is more factual and shows a fairly different positioning than the other four organizations. AWF positions itself more as a knowledge provider. It does not focus as much as WWF or TNC on developing a stronger bond between nature and human beings. It does not focus on denouncing environmental damages. Rather, it raises awareness about the necessity to protect nature through knowledge and information. Hence, it conducts more community building tactics.

With 102 Facebook posts over the one-month period, AWF is the second most active organization of this sample. Seventy-six posts out of 102 were about providing information. This corresponds to 74% of all its published content. The advocacy strategy is to provide information about nature conservation, facts and figures, and the work of the organization, and to showcase the beauty of nature in Africa (e.g. on 30 June: "Rhinos have a very good sense of smell and hearing although their eyesight isn't the best. This explains why rhinos will occasionally be seen charging for no reason. https://bit.ly/2IvPcbo").[77] The second advocacy tactic concerns community. In this case, AWF focuses on quizzes and photos, asking the audience for example to propose a caption for a photo published by the organization (e.g. on 26 June: "Help us #caption this photo!" or on 25 June: "Are you as smart as an elephant? As social as a chimpanzee? Take AWF's quiz to see what your African spirit animal is! http://bit.ly/1KiaRxy").[78] Six out of 102 posts are considered calls for action, and they focus on two elements: fundraising and pledges (e.g. on 27 June: "Africa is expected to bear the brunt of the adverse effects of climate change, including a drastic decrease in water resources. Pledge to live more sustainably today! www.awf.org/fight-climate-change").[79] The most shared post was published on 21 June with a community tactic that wished a "Happy Giraffe Day" to its audience.

With 135 tweets, AWF is also quite active on Twitter. Similar to its usage of Facebook, its main advocacy tactic is by far information (78%). Although it provides content about the work of the organization (e.g. on 15 June: "African Wildlife Foundation's mission is to ensure wildlife and wild lands thrive in modern Africa. https://bit.ly/2FMqwaP"),[80] the main content is about species in Africa: what their extinction state is, facts and figures about them (e.g. on 3 June: "#Baboons are some of the largest monkeys in the world. The baboon weighs between 9 to 31 kilograms (20 to 70 pounds). At the shoulder, they vary between 50 to 76 centimeters (20 to 30 inches). https://bit.ly/2IG59fM").[81] The tone of voice used is more formal than that of other organizations from this sample (Ocean Conservancy for instance), and the story line is quite factual, focusing on raising awareness about nature (and the state of nature) in Africa. In terms of community tactics, AWF primarily asks its audience to suggest a caption for a photo and to take part in a quiz (e.g. on 9 June: "Pop quiz: Is this a rhino? http://bit.ly/2JpSPNd").[82] In terms of action, AWF mobilizes its community to raise funds (e.g. on 4 June: "Help African Wildlife Foundation every time you shop: www.iGive.com/P4cvtCa #iGiveDoYou")[83] and to sign a petition (e.g. "Tell #Congress to pass The #Wildlife Conservation and Anti-Trafficking Act of 2018! www.awf.org/support-hr5697").[84] The tweet with the most likes was published on 27 June. It illustrates the fairly official tone of voice that AWF uses to celebrate good news: "There is hope for #Africa's mountain #gorillas. In the 1980s, there were about 600 of these great apes. Today, there are more than 1,000. http://bit.ly/2sqNlKL."[85]

The last organization of this sample has a topical scope: Ocean Conservancy. Its story line showcases the beauty of oceans and the need to protect them. Similarly to WWF and TNC, the organization focuses its advocacy strategies on developing a stronger bond between human beings and nature (in this case oceans). Its audience seems also to be the general public. However, the tone of voice is quite different than that of WWF or TNC. The organization makes more extensive use of humor to touch its audience. Its tone of voice is more informal and friendly, addressing the audience directly.

On Facebook, Ocean Conservancy's publication rate is within the average, with 63 posts published on Facebook during this one-month period. Its advocacy tactics are divided mainly between information (50%) and community (42%). It is the only organization in this sample that regularly uses humor in its tweets (e.g. on 1 June: "You have got to be squidding me. 🐙 635 likes").[86] In terms of information, the organization raises awareness about the state of marine species (e.g. on 29 June: "This microbiologist studies large filter feeding fish, mostly mobulid rays and whale sharks. Find out how microplastics are affecting these majestic ocean creatures. http://ow.ly/1Ugb30kCPEO"),[87] coasts and beaches, as well as some domestic political issues (e.g. on 20 June: "We will hold the Administration accountable over the #NationalOceanPolicy – America is an ocean nation. We are committed to coastal communities and a healthy ocean, both during #NationalOceanMonth and every day. http://ow.ly/QHcC30kAJQp").[88] In terms of community, Ocean Conservancy organized a photo contest (e.g. on 25 June: "Summertime is officially here and among the many incredible things it brings along with it is the annual Ocean Conservancy Photo Contest! Whether you have coastal photos from previous vacations or underwater pictures from a stellar scuba session, we want to see them entered into our Annual Photo Contest! http://ow.ly/gGBm30kEVhv"),[89] but it is mainly the friendly and optimistic posts that feature nature to empower Facebook users in their everyday lives that catch their attention (e.g. on 29 June: "Rough week? It's Friday . . . have a laugh! 😄 http://ow.ly/bgvx30kCTUi").[90] In terms of action, the organization mobilizes its audience to sign petitions, collect waste in coastal areas, and raise funds (e.g. on 19 June: "They're here! Nathan Quinn and his Force Blue Teammates have officially welcomed 6 new special operations veterans to their underwater force. You can meet the recruits by clicking the link below; plus, learn how YOU can help sponsor one of these Special Operations Veterans throughout their training! #OneTeamOneFight #NationalOceanMonth #ForceBlue Meet Force Blue Team II >>> www.ForceBlueTeam.org <<<").[91] The post with the most likes was published on 6 June and illustrates well the friendly and informal tone of voice of the organization: "One beluga is already awesome enough . . . but 600? 🐋 Pardon us, be back soon, we believe there's a #WhaleWednesday party taking place that we need to attend. 🐳 How much do you know about belugas? Find out with our Wildlife Fact Sheets! → https://bit.ly/2jZuzqb (Video: https://bit.ly/2IiH9zg)."[92]

With 171 tweets, Ocean Conservancy is also active on Twitter. Its advocacy strategy however is quite different from that of the other organizations, with far more community building tweets (47%). Also, the general tone of voice is more friendly, and many tweets impersonate nature and species (e.g. on 3 June: "As my name suggests, you wouldn't want to put me in a china shop. I can grow to 12 feet long & 500 pounds. But I am agile, despite my size! I've been spotted leaping up river rapids, almost like salmon, en route to inland lakes. #BullShark #Shark-Sunday http://ow.ly/FYOt30kiDIQ").[93] Many tweets include positive quotes from environmentalists and poets (e.g. on 10 June: " 'Live in the sunshine. Swim in the sea. Drink in the wild air.' – Ralph Waldo Emerson #OceanOptimism #NationalOceanMonth #SeaLionSaturday http://ow.ly/wOgP30kkn45").[94] Many embrace an optimistic view and aim to portray nature as empowering and motivating the audience for their everyday life (e.g. on 8 June: "We all have that one friend we love more than all our 'otter' friends. Make sure you let them know today how much you care! #NationalBestFriendDay #WorldOceansDay #OtterFriends http://ow.ly/IonB30knbae"[95] or on 29 June: "Keep your friends close and your anemones closer. #FishyFriday http://ow.ly/xPl030kHdKQ").[96] There was also a photo contest taking place during this month. Furthermore, many tweets directly engage some members of their audience and thank them for their actions (e.g. on 23 June: "Thank you @ChrisCoons for backing strong funding for @NOAA ocean programs that support sustainable environment and economic development across the country! #TheMoreYouNOAA").[97] In terms of information, Ocean Conservancy focuses mainly on species and domestic protection laws. In terms of action, the focus is mainly domestic, asking the audience to take part in adopting more sustainable behaviors and making their voice heard (e.g. on 5 June: "Will you speak up for the #ESA? Let your Members of Congress know that you support the critical protections the Endangered Species Act provides our nation's most vulnerable marine animals. #WorldEnvironmentDay http://ow.ly/SIQk30kiach").[98] The tweet with the most likes clearly illustrates the advocacy tactics used by Ocean Conservancy on Twitter. It was published on 12 June: "A journey of a thousand miles begins with a single step #TurtleTuesday #OceanOptimism #NationalOceanMonth http://ow.ly/ML4i30kknBs."[99]

7. Concluding remarks

This chapter confirms the dominant narrative that hints that social media instruments have become an integral part of the advocacy strategies of CSOs. The increasing use of ICTs and in particular social media in civic interactions is probably the most debated civil society phenomenon in recent years.[100] Social media platforms enable CSOs not only to reach out to a wider community, but also to raise awareness about the protection of the environment. Hence, advocacy is one of the key roles of CSOs on the international stage that is most affected by the generalization of social media.

NGOs accredited to UNEA use social media to set the agenda according to their interests and objectives. Their advocacy tactics depend largely on their mission. The choice of social media tactics, tone of voice, regularity of publishing, and story line depend on how they are positioned in global environmental governance. Their communications were coherent throughout the month and clearly illustrate the variety of advocacy strategies that are possible thanks to social media platforms. Notwithstanding, the tactics do not depend on the medium (Facebook or Twitter), the geographical scope (regional or international), or the topical scope (global or marine).

However, the five organizations examined in this chapter showed first informative rather than consultative and mobilizing advocacy tactics, no matter the organization, its audience, and geographic or thematic scope. As Lovejoy argues: social media tends to be used "as an extension of information-heavy websites. These organizations are missing the bigger picture of its uses as a community-building and mobilizational tool."[101]

Moreover, a large majority of the environmental NGOs accredited to UNEA are present on Facebook and Twitter. However, the digital divide remains an issue, with the number of NGOs from Europe and North America surpassing African ones by far.

Lastly, this analysis also showed how a handful of environmental NGOs concentrate the attention of social media. These organizations have successfully developed a large audience, and concentrate the capacity to attract the attention of Facebook and Twitter users. This implies that they have the strongest influence on social media platforms, and therefore are the best suited to frame environmental issues and discourses and raise awareness based on their views and values. Hence, it shows that social media platforms have not only allowed a large number of organizations to reach out to global audiences, but also contributed to a concentration of the capacity to capture the attention of citizens. Hence, social media platforms tend to inhibit smaller organizations from raising awareness about other issues and proposing other perspectives about the protection of the environment. Therefore social media platforms tend to reduce the plurality of sources of information for citizens.

The use of social media platforms also sheds light on the participation capacity and strategies of CSOs on the international stage. Indeed, their advocacy strategies allow them to participate in global environmental governance by making their voices heard, and by becoming new forms of authority, and in that context support SDG Target 16.6, "Develop effective, accountable and transparent institutions at all levels,"[102] and SDG Target 16.7, "Ensure responsive, inclusive, participatory and representative decision-making at all levels." Their use of social media platforms to inform, consult with, and mobilize their audience also enables them to gain more visibility on the international stage, and by doing so, to better cooperate with other actors and help achieve SDG Goal 17 entitled "Revitalize the global partnership for sustainable development."[103] However, as mentioned previously, the concentration of audience in the hands of a small number of organizations contrasts with these assumptions. On one hand, social media platforms support many civil society organizations in gaining visibility and reaching out to a larger audience. On the other hand, this research showed that access to a larger audience is limited to the biggest and oldest environmental NGOs. Hence, social media platforms reaffirm the inequalities and participation patterns that exist offline.

Notes

1 The terms "social media," "social media platforms," and "social media networks" will be used interchangeably to describe the same reality.
2 Stevens, T.M., Aarts, N., Termeer, C.J.A.M., Dewulf, A., 2016. Social media as a new playing field for the governance of agro-food sustainability. *Current Opinion in Environmental Sustainability*, 18, p. 99. http://dx.doi.org/10.1016/j.cosust.2015.11.010
3 Ibid.
4 Soanes, C., Stevenson, A., 2004. *Concise Oxford English dictionary*, 11th ed. New York, NY: Oxford University Press, p. 19.
5 Lovejoy, K., Saxton, G.D., 2012. Information, community, and action: How nonprofit organizations use social media. *Journal of Computer-Mediated Communication*, 17(3), p. 341. doi:10.1111/j.1083-6101.2012.01576.x
6 Guo, C., Saxton, G., 2012. Tweeting social change: How social media are changing nonprofit advocacy. *Nonprofit and Voluntary Sector Quarterly*, 41, p. 61.
7 OECD, 2001. *Citizens as partners: OECD handbook on information, consultation and public participation in policy-making*. Paris: OECD Publications Service, pp. 15–16.
8 Goal 16. *Promote peaceful and inclusive societies for sustainable development, provide access to justice for all and build effective, accountable and inclusive institutions at all levels.* See more here https://unstats.un.org/sdgs/indicators/indicators-list/
9 SDG Target 16.6. *Develop effective, accountable and transparent institutions at all levels.* See more here https://unstats.un.org/sdgs/indicators/indicators-list/
10 Xprize. *Op Cit.*
11 Statistica, 2017. www.statista.com/statistics/278414/number-of-worldwide-social-network-users/
12 Goldkind, L., 2015. Social media & social service: Are nonprofits plugged in to the digital age. *Human Services Organizations: Management, Leadership & Governance*, 39(4), p. 380.
13 Lovejoy, K., Saxton, G.D., 2012. Information, community, and action: How nonprofit organizations use social media. *Journal of Computer-Mediated Communication*, 17(3), p. 338. doi:10.1111/j.1083-6101.2012.01576.x
14 Roback, A.J., 2013. Uncovering motives for social networking site use among practitioners at non-profit organizations. *Proceedings of the 2013 conference on computer supported cooperative work companion*, ACM, New York, NY, p. 78.
15 Bortree, D., Seltzer, T., 2009. Dialogic strategies and outcomes: An analysis of environmental advocacy groups' Facebook profiles. *Public Relations Review*, 35, p. 318. doi:10.1016/j.pubrev.2009.05.002
16 Lovejoy, K., Saxton, G.D., 2012. *Op Cit.*, p. 351. doi:10.1111/j.1083-6101.2012.01576.x
17 Madway, G., 2010. Twitter remakes website, adds new features. *Reuters.* www.reuters.com/article/idUSN1411135520100915
18 Valenzuela, S., 2013. Unpacking the use of social media for protest behavior: The roles of information opinion expression, and activism. *American Behavioral Scientist*, 57, p. 923.
19 Adamoli, G., 2012. Social media and social movement's: A critical analysis of audience's use of facebook to advocate food activism offline. *Electronic Theses, Treatises and Dissertations*, p. 3.
20 Bennett, W.L., Segerberg, A., 2011. Digital media and the personalization of collective action. *Information, Communication & Society*, 14, p. 771.
21 Oxford Dictionary. https://en.oxforddictionaries.com/definition/advocacy
22 Lovejoy, K., Saxton, G.D., 2012. *Op Cit.*, p. 341.
23 Ibid.
24 Ibid.
25 Guo, C., Saxton, G., 2012. *Op Cit.*, p. 71.

26 Ibid.
27 Ibid.
28 FitzGerald, E., McNutt, J.G., 1999. Electronic advocacy in policy practice: A frame-work for teaching technologically based practice. *Journal of Social Work Education*, 35(3), p. 335.
29 Pew Internet Research. www.pewinternet.org/2016/11/11/social-media-update-2016/
30 Guo, C., Saxton, G., 2014. Tweeting social change: How social media are changing nonprofit advocacy. *Nonprofit and Voluntary Sector Quarterly*, 41(1), p. 58.
31 Tredinnick, L., 2006. Web 2.0 and business: A pointer to the intranets of the future? *Business Information Review*, 23(4), p. 230. doi:10.1177/0266382106072239
32 Goldkind, L., 2015. *Op Cit.*, p. 383.
33 Stieglitz, S., Dang-Xuan, L., 2013. Emotions and information diffusion in social media – sentiment of microblogs and sharing behavior. *Journal of Management of Information Systems*, 29, p. 241.
34 Vasterman, Peter, Joris Yzermans, C., Dirkzwager, Anja J.E., 2005. The role of the media and media hypes in the aftermath of disasters. *Epidemiologic Reviews*, 27(1), p. 111. https://doi.org/10.1093/epirev/mxi002
35 The terms "CSOs" and "NGOs" will be used indistinctively to describe the same group of actors. "NGOs" is used by UN Environment on its website.
36 UNEA. http://web.unep.org/environmentassembly/node/40734
37 The number of accredited non-governmental organizations changes over time. This analysis was based on the number of organizations published on the UN Environment website in June 2018.
38 UNEA.http://web.unep.org/about/majorgroups/resources/list-accredited-organizations
39 UNEA.   http://web.unep.org/about/majorgroups/modalities/accreditation   [Accessed June 2018].
40 UN   Environment   Accreditation   criteria.   www.unenvironment.org/civil-society-engagement/accreditation
41 Guo, C., Saxton, G., 2012. *Op Cit.*, p. 61.
42 The top 15 environmental NGOs accredited to UNEA, whose main mission is dedi-cated to the protection of the environment (not poverty reduction or humanitarian aid for instance), and with the most users liking their official page.
43 Greenpeace International. www.facebook.com/pg/greenpeace.international/about/
44 Sierra Club. www.facebook.com/pg/SierraClub/about/
45 NRDC. www.facebook.com/pg/nrdc.org/about/
46 IFAW Deutschland. *(. . .) beispielsweise mit Kampagnen gegen die kommerzielle Jagd auf Wale und Robben.* www.facebook.com/pg/ifaw.de/about/
47 WWF. www.facebook.com/pg/WWF/about/
48 Ibid.
49 Nature Conservancy. www.facebook.com/pg/thenatureconservancy/about/
50 African   Wildlife   Foundation.   www.facebook.com/pg/AfricanWildlifeFoundation/about/
51 Ocean Conservancy. www.facebook.com/pg/oceanconservancy/about/
52 Number of users who liked their official pages. As of 6 July 2018.
53 Twitter. https://help.twitter.com/en/using-twitter/twitter-lists
54 Twitter. https://help.twitter.com/en/using-twitter/twitter-moments
55 The top 15 environmental NGOs accredited to UNEA, whose main mission is dedi-cated to the protection of the environment (not poverty reduction or humanitarian aid for instance), and with the most users liking their official page.
56 Number of users who liked their official pages. As of 6 July 2018.
57 Although TEMA has a large number of followers, its tweets are in Turkish (a language foreign to the author). This means that their audience is mainly domestic and that the analysis of its advocacy tactics would have required translation. Therefore the choice was made to disqualify it for this part of the analysis.

58 Excluding organizations such as ICRC and IFRC, which are not environmental organizations stricto sensu, and TEMA, which publishes in Turkish.
59 Guo, C., Saxton, G., 2012. *Op Cit.*, p. 74.
60 WWF. www.facebook.com/pg/WWF/posts/
61 Ibid.
62 WWF. https://twitter.com/wwf
63 Ibid.
64 Ibid.
65 Ibid.
66 Greenpeace. https://twitter.com/greenpeace
67 Ibid.
68 Ibid.
69 The Nature Conservancy. www.facebook.com/pg/thenatureconservancy/posts/
70 Ibid.
71 Ibid.
72 The Nature Conservancy. https://twitter.com/nature_org
73 Ibid.
74 Ibid.
75 Ibid.
76 Ibid.
77 African Wildlife Foundation. www.facebook.com/pg/AfricanWildlifeFoundation/posts/
78 Ibid.
79 Ibid.
80 African Wildlife Foundation. https://twitter.com/AWF_Official
81 Ibid.
82 Ibid.
83 Ibid.
84 Ibid.
85 Ibid.
86 Ocean Conservancy. www.facebook.com/oceanconservancy/
87 Ibid.
88 Ibid.
89 Ibid.
90 Ibid.
91 Ibid.
92 Ibid.
93 Ibid.
94 Ibid.
95 Ibid.
96 Ibid.
97 Ibid.
98 Ibid.
99 Ibid.
100 Edwards, M., 2014. *Civil society*. Cambridge, UK: Polity Press, p. viii.
101 Lovejoy, K., Saxton, G.D., 2012. Information, community, and action: How non-profit organizations use social media. *Journal of Computer-Mediated Communication*, 17(3), p. 341. doi:10.1111/j.1083-6101.2012.01576.x
102 SDG Target 16.6. *Develop effective, accountable and transparent institutions at all levels*. See more here https://unstats.un.org/sdgs/indicators/indicators-list/
103 Xprize. *Op Cit.*

4   Participation and the digitalization of a multi-stakeholder governance process

1. Introduction

Founded in 1948, the International Union for Conservation of Nature (IUCN) is one of the oldest global environmental organizations. It was founded by 18 governments, seven international organizations, and 107 national nature conservation organizations, which all agreed to sign the Constitutive Act founding an International Union for the Protection of Nature.[1] It is an environmental network of more than 10,000 experts in six commissions dedicated to species survival, environmental law, protected areas, social and economic policy, ecosystem management, and education and communication, and whose members produce high-level scientific knowledge in the field of nature conservation. It is also a membership organization with 1,300 Member organizations coming from civil society, indigenous groups, and governmental agencies and their representatives. They provide good representation of the environmental community at large. IUCN's staff, located in Switzerland and in 50 offices throughout the world, supports its Members by producing scientific knowledge, advancing their views on the international stage, facilitating collaboration for conservation field projects, and providing a global forum for governmental and non-governmental actors to discuss human progress, economic development, and nature conservation. Members can join National Committees that consist of IUCN Members located within a specific state and region. These committees help coordinate the participation of Members in IUCN's governance processes and enhance collaboration among them.[2]

One of the key features of the organization is its membership services, and more precisely its multi-stakeholder processes that give voice to a wide range of environmental actors. This – almost unique – multi-stakeholder process is at the heart of the organization and enables Members to discuss the program of the organization, namely work objectives for each four-year period, vote on the budget, elect management bodies, and make proposals about the protection of the environment and governance processes. IUCN is one of the only organizations in the world to provide an official status to non-governmental actors equal to state Members. For decades, non-governmental Members have been able to vote on and participate in core governance processes on an equal footing with governmental actors. This is truly remarkable, and should justify more public and research interest in this organization.

Since 1948, IUCN has organized 25 General Assemblies and World Conservation Congresses. IUCN had General Assemblies until the Montreal Congress in 1996; afterwards they have been called World Conservation Congresses. In its most recent format, each Congress consists of a Forum where IUCN constituents meet people from all around the world to discuss issues related to nature conservation and sustainability challenges.[3] It consists of a number of events categorized per topic, and includes workshops, panels, posters, high-level dialogues, and social events. Following the Forum, IUCN membership gathers for a Members' Assembly to address questions pertaining to IUCN membership including voting on the program, motions, and electing the next Council. It allows IUCN Members "to consider major thematic areas of the IUCN Programme and to facilitate the sharing of information and experience."[4] Furthermore, according to article 20 (a) and (b) of the Statutes, the objectives of Congress are "(a) to define the general policy of IUCN; (b) to make recommendations to governments and to national and international organizations in any matter related to the objectives of IUCN."[5]

IUCN enables its Members to vote on a new program and budget, and elect the new governance body (Council). But not only this. Beyond these traditional governance procedures, IUCN has developed the motions process. Prior to the Congress, Members can make proposals that will be discussed on site. Each proposal, called a motion, is the result of collaboration between several Members: the one who makes the proposal, called the sponsor, and the others who support the proposal, called the co-sponsors. Traditionally, meaning before the generalization of information and communication technologies (ICTs), IUCN Members would coordinate through phone and in-person meetings and then send the proposal to headquarters. The motion would then be discussed at Congress by a large number of groups made up of both governmental and non-governmental Members. Once the text was agreed, the final version would be submitted to the Plenary for a vote. If accepted, it became a Resolution for the organization to execute, or a Recommendation for external stakeholders. If refused, the text went back to groups for debate, and if no agreement was found, the motion was finally rejected.

So far, IUCN's Members have adopted more than 1300 Resolutions on a wide range of issues including climate change, species, protected areas, marine, and water, to name only a few.[6] This process has made possible the adoption of several international environmental instruments, standards, and agreements.[7] Among other accomplishments, IUCN helped set up the World Wide Fund for Nature (1961), created the World Heritage List with UNESCO (1972), helped establish the Convention on International Trade in Endangered Species (1973), drove the establishment of the Convention on International Trade in Endangered Species of Wild Fauna and Flora (1973), coined the term "sustainable development" (1980), proposed the Convention on Biological Diversity (1982), and helped drive the Rio Earth Summit (1992). During the last Congress, an IUCN Resolution led to the establishment of the Global Judicial Institute for the Environment.

This chapter will focus on the motions process and more precisely its progressive digitalization, which sheds light on the impact of ICTs on the participation of state and non-state actors in a multi-stakeholder governance process. Thanks to digital technologies, CSOs, which traditionally have fewer resources than states, can gain additional visibility and take a stronger role in these governance processes, which supports SDG Target 16.6, "Develop effective, accountable and transparent institutions at all levels,"[8] and SDG Target 16.7, "Ensure responsive, inclusive, participatory and representative decision-making at all levels." Their use of ICTs also allows them to better cooperate with other actors and help achieve SDG Goal 17 entitled "Revitalize the global partnership for sustainable development."[9]

This chapter will first discuss the process in detail, and then analyze the use of ICTs and their impact on the participation of various governmental and nongovernmental stakeholders. The opportunity to evaluate how ICTs affect the role played by these two types of actors is quite unique and should illustrate the potential of digital technologies when it comes to improving representative decision-making processes on the international stage.

2. Participation in the motions process

Participation in decision-making processes is a vast topic that keeps many professionals and scholars busy. According to the Cambridge Dictionary, inclusiveness means "the quality of including many different types of people and treating them all fairly and equally."[10] On the international stage, this implies having a clear view of the different actors and providing a space for them to have a say in the policy-making processes that affect them. In the case of IUCN, inclusiveness is pretty straightforward, since it concerns the Members of the organization. A representative process implies that all types of actors have access to governance processes on an equal footing. IUCN counts four categories of Members: (A) states and governmental agencies (including political and economic integration organizations), (B) national and international NGOs, (C) indigenous peoples' organizations, and (D) affiliates.[11] Membership comprises 85+ states, 120+ governmental agencies, 1000+ non-governmental organizations, and 15+ indigenous groups.[12] The C category was formally created at the 2016 Congress. The objective of inclusive participation therefore implies that all types of Members take part in the governance process, and that their participation is representative of their importance among IUCN's membership: inclusive and representative participation.

This definition of participation is aligned with SDG 16, "Promote peaceful and inclusive societies for sustainable development, provide access to justice for all and build effective, accountable and inclusive institutions at all levels," and more specifically with Target 16.7, "Ensure responsive, inclusive, participatory and representative decision-making at all levels" and Target 16.8, "Broaden and strengthen the participation of developing countries in the institutions of global governance." However, these two targets must be adapted to IUCN's specific membership types. This means that inclusive and representative participation within IUCN excludes the private sector; multinational corporations (MNCs) and small and medium enterprises (SMEs) can collaborate with IUCN on projects but cannot become Members. Due to its impact on the environment, the private sector is a major actor in global environmental governance. The fact that it is excluded from IUCN's membership could be perceived as a negative point. However, its inclusion could dramatically change the power play among actors and the profile of the organization, and hence is a complex decision to make.

IUCN's motions process is a concrete example of the inclusive and representative participation that Members enjoy in the decision making of the organization. According to the IUCN Statutes and Regulations, the process starts with a motion, which is defined as "a draft in writing of any decision which the World Congress is requested to take. Such motion may take the form of a resolution, recommendation, expression of opinion or proposal. Resolutions are directed to IUCN itself. Recommendations are directed to third parties, and may deal with any matter of importance to the objectives of IUCN."[13] The objective of motions is therefore to outline the general policy of the organization, have an impact on the policies and initiatives of third parties, and tackle governance issues.[14] Motions can be proposed by the Council or eligible Members (all categories of Members except affiliate Members),[15] and each motion must be supported by at least five other eligible Members, which are called co-sponsors, as mentioned previously. When necessary, an explanatory memorandum can be appended to the text of the motion and circulated among Members.[16] Commission members and IUCN staff cannot propose motions or vote. However, IUCN staff assist Members in submitting motions, and provide technical advice to the Council.

Motions are submitted prior to each Congress (a minimum of six months before) and sent to the Director General. In preparation for Congress, a Motions Working Group is set up by Council and comprises Councilors and Member representatives who oversee the motions process. More precisely, the Working Group "(1) establishes specific procedures for the motions process and principles upon which they will be revised; (2) ensures that the statutory requirements are applied to the submitted motions and that motions are treated fairly and equitably; (3) advises sponsors of motions to revise, amend or withdraw a draft motion; (4) prepares motions for tabling at Congress; and (5) facilitates discussion between Members on motions in advance of the Congress."[17]

If a motion is presented at a later stage, meaning directly at Congress, special rules apply: it must be submitted either by the Council or by a sponsor with a minimum of ten co-sponsors all eligible to vote, and, in agreement with the Resolutions Committee,[18] the topic of the motion must be new, urgent, and unforeseen.[19]

Motions submitted must abide by certain rules in terms of both content and format, and the Motions Working Group will verify this before accepting them as described in article 54[20] of the IUCN Statutes and Regulations (revised prior to the 2016 Congress). The conditions to submit a motion were previously less strict: they had to be consistent with the objectives of IUCN and could not repeat a prior decision unless the issue had not been resolved. Indeed, during the 2012 Congress in Jeju, Republic of Korea, IUCN Members adopted Resolution

WCC-2012-Res-001 entitled "Strengthening the motions process and enhancing implementation of IUCN Resolutions" in which they requested the Director General to revise the motions process and in particular "to address the issue of the increasing number of motions."[21] Further to this Resolution, the Motions Advisory Group worked with the Secretariat to propose a revision of the motions process, wherein a majority of motions are debated and voted online prior to the onsite Congress.[22] Article 54 of the IUCN Statutes and Regulations that describes the conditions for a motion to be accepted by the Motions Working Group was then amended. The new conditions are explained below.

In terms of content, motions must be consistent with the IUCN objectives "to influence, encourage and assist societies throughout the world to conserve the integrity and diversity of nature and to ensure that any use of natural resources is equitable and ecologically sustainable."[23] Their goals, along with the means required by Members, if called for, must be reasonable and achievable, and they should not repeat the content of previous resolutions or recommendations. Finally, they must prove that when the issue is at the local, national, and/or regional level, it has already been tackled unsuccessfully by the relevant instances.[24]

In terms of process and format, motions must be submitted before the deadline. They must be proposed and co-sponsored by Members eligible to vote (not affiliate Members), and sponsors and co-sponsors must be in good standing, meaning that their dues are paid at the time of the submission of motions.[25] The main sponsor must specify which Members are called upon for action and what resources are necessary to implement the motion. When motions address an issue outside of the state or region of the sponsor, at least one co-sponsor must belong to the state or region concerned. Finally, the template approved by Council must be used.[26]

Once the motion is submitted, only Members from categories A, B, and C can cast a vote either online or at Congress (depending on the motion). Votes from Members of categories B and C are counted together to produce one single combined vote.[27] Decisions at Congress are taken by a simple majority of votes (unless stated differently in the Statutes), and abstentions do not count as votes cast. Each category of Member has distinct voting rights. Category A Members (governmental Members) have three votes. If a state is a Member of IUCN and also has governmental agencies that are Members of IUCN, one of its three votes must be exercised collectively by the government agencies. If governmental agencies are Members of IUCN but their state of origin is not, they collectively (in the case of more than one) have one vote. If Member states belong to a political or economic organization, and this organization is a Member of IUCN, they can together choose how to distribute votes, but the total number of votes cannot exceed those of all Member states that are part of this organization.[28] In addition, category B Members have either one vote (in the case of non-governmental organizations) or two votes (in the case of international non-governmental organizations).[29] Lastly, category C Members have one vote. Indigenous peoples' organizations gained this new status at the 2016 Congress, and therefore will be eligible to vote only at the next Congress in 2020 in France.[30]

As mentioned previously, the evolution of IUCN's membership follows the global trends mentioned in Chapter 1. Its membership grew exponentially in the 1990s and has slowed down in the 2000s. Any change in participation in the motions process after the year 2000 can therefore not be explained solely by an increase in the number of Members. As of June 2018, IUCN counted 1344 Members.[31] Forty-one are affiliates, which means that they do not have any voting rights. One hundred twenty-seven are government agencies, and 89 are states (category A). One hundred seven are international NGOs and 963 national NGOs (category B). Lastly, the most recent category (C), indigenous peoples' organizations, counts 17 entities. In other words, category A accounts for about 16%, category B for about 80%, category C for about 1%, and affiliates for about 3% of total membership. If we discount affiliates that cannot vote, states represent about 7%, governmental agencies about 10%, international NGOs about 8%, national NGOs about 74%, and indigenous peoples about 1% of all Members that can cast a vote. In comparison, IUCN counted 1143 Members in 2011, with a similar distribution among categories.[32]

Interestingly, national NGOs, which are often considered marginalized actors on the international stage due to their limited resources, represent the vast majority of IUCN membership. These Members are the first ones to benefit from the skills revolution, where the use of ICTs has granted them better access to information, and enabled them to better collaborate and participate in global decision-making processes, as discussed in the first chapter of this book. However, for a motion to be adopted, a simple majority of Members of category A and of categories B and C together must be met. This rule ensures that all types of Members have a say and their voices are heard, even if their group represents a small portion of all Members (category A to be precise).

3. Progressive use of ICTs to submit a motion

Thanks to the progressive generalization of ICTs, IUCN Members gained an additional capacity to submit a motion. As discussed in this section, this new capacity led them to consider and adopt a larger number of recommendations and resolutions over time.

The IUCN Resolutions and Recommendations platform on IUCN's portal allows its users to search through resolutions and recommendations (motions that were adopted) by criteria (single or multiple simultaneously). It is accessible to everyone, thereby enhancing the transparency and visibility of the Union as a whole.[33] Each resolution and recommendation has been given a specific code, related key words, and geographical scope. This enables users to search by code and keyword, but also by title, country, region, type, conference, and free text. It is also possible to determine the implementation status of each resolution and recommendation through the online tracking system.

As of June 2018, the IUCN Resolutions and Recommendations platform counted 1305 entries (resolution or recommendation). Eight hundred seventy-one had a global scope (about 67%), 895 were resolutions (about 69%), and 410 were recommendations (about 31%). The main topics covered by resolutions and recommendations since 1948 are as follows:

- Species (393 resolutions and recommendations),
- Protected areas (350 resolutions and recommendations),
- International agreements and processes (339 resolutions and recommendations),
- Marine (243 resolutions and recommendations),
- Education, capacity building, public awareness, and communication (237 resolutions and recommendations),
- Ecosystems (229 resolutions and recommendations),
- Environmental governance (223 resolutions and recommendations),
- Human well-being (205 resolutions and recommendations).

These results highlight IUCN's engagement for the protection of species and ecosystems in the world. The IUCN Red List of Threatened Species[TM34] and the IUCN Global Protected Areas Programme[35] are some well-known illustrations of this mission. These topics are widely communicated; the general public is quite aware of these questions through traditional and digital media, with regular reports on the web, television, radio, and in newspapers.

Furthermore, as mentioned previously, IUCN's resolutions and recommendations process has led to the creation of several environmental organizations and to the development of new conventions and treaties. This is demonstrated with 339 resolutions and recommendations about international agreements and processes and 223 about environmental governance (together reaching 562). Also, resolutions and recommendations on education and human well-being indicate that the organization over the years has adopted a certain consensus on what environmental protection entails and IUCN Members share a common understanding of it. These topics are reflected in the organization's mission: "Influence, encourage and assist societies throughout the world to conserve the integrity and diversity of nature and to ensure that any use of natural resources is equitable and ecologically sustainable."[36] They also illustrate well the three pillars of IUCN's work – nature conservation, environmental governance, and scientific knowledge.

In addition, the number of adopted resolutions and recommendations that concern "education, capacity building, raising awareness, and communication" indicates how important this theme has become for the environmental community and global environmental governance. To some extent, "education, capacity building, raising awareness, and communication" could be translated as new power and improved participation in international governance through education, capacity building, raising awareness, and communication.

In terms of time range, IUCN membership adopted the first recommendation in 1988 during the San Jose General Assembly. This came after the 1984 Congress in Madrid where Members asked the organization to "develop procedures to provide input into international activities in cooperation with relevant

IUCN components."[37] Furthermore, it followed the officialization of the concept of "sustainable development" by the Brundtland Commission in its 1987 report "Our Common Future." Two of the recommendations adopted in 1988 illustrate well how the resolutions and recommendations process is fully integrated into global environmental governance debates and challenges. One example is Recommendation GA 1988 RES 026 entitled "Report of the World Commission on Environment and Development" that recognizes the worldwide "consensus on the nature of sustainable development" and "urges governments, international bodies, nongovernmental organizations, and individuals to examine the Commission's Report, consider its recommendations, and join in efforts to solve the problems it identifies."[38]

On average, since 1988, Members have adopted 45 recommendations per Congress. The Congress with the highest number was in 1996 with 71 recommendations. In terms of resolutions, the average number adopted per Congress since the creation of IUCN has been about 36. However, between 1996 and 2004, the number doubled from 40 to 80, and continued to increase in 2008 with 106 and in 2012 with 137. In other words, the number of resolutions was multiplied by 3.4 between 1996 and 2012 and doubled between 2000 and 2012. The conditions for submitting motions were identical for the Congresses of 2004, 2008, and 2012.[39] This explosion in the number of resolutions ceased after 2012 for a specific reason: the tightening up of the conditions to submit a motion (art. 54 of the IUCN Statutes and Regulations). Indeed, Members asked the Union to reduce the number of motions after 2012 and to develop a more efficient and precise process to submit, discuss, and vote on motions. The objective was also to offer more time during the Members' Assembly for Members to debate on more controversial issues. The new version of the motions process was applied for the first time in 2016. Due to this change, the motions process will be examined for two timeframes: prior to and since 2016.

Looking at the period 2000–2012, a sharp increase in the number of resolutions can be noted. In other words, since the generalization of new ICTs, IUCN Members have adopted about four times more global resolutions and recommendations. As mentioned previously, this change cannot be explained by an increase in IUCN's membership. The fact that IUCN Members have adopted four times more resolutions and recommendations since 2000 could be possible thanks only to the use of ICTs: the use of websites to access information, and emails, text messages, and Skype to collaborate with other Members on the elaboration of new motions. IUCN Members have access to more information and can communicate more intensively with each other prior to World Conservation Congresses, leading to an increase in the number of motions submitted and thus resolutions and recommendations adopted.

Given that this research focuses on global environmental governance, only resolutions and recommendations with a global geographical scope were chosen here: out of 1,193 resolutions and recommendations adopted between 1948 and 2012, 785 have a global outreach (66%).[40] Four hundred thirty-one were adopted before 2000, and 354 were adopted afterwards. To be more precise, 431 resolutions and recommendations with a global scope were adopted during the 20 Congresses before 2000 (55%), covering 52 years, and 354 were adopted during the four Congresses after 2000 (45%), covering only 12 years. Per event, IUCN Members adopted an average of 21.55 resolutions and recommendations with a global scope before 2000, and 88.5 since 2000.

IUCN's regions are specific to the organization's internal governance; North America is associated with the Caribbean for instance. The results sorted by region are therefore specific to IUCN and cannot be extrapolated for general consideration. However, if taken region by region, the results of the search nevertheless show a similar increase in the number of resolutions and recommendations adopted. On average, they increased fourfold. As Table 4.1 indicates, IUCN Members adopted 37 resolutions and recommendations about Africa before 2000 (1.85 per Congress) and 30 after (7.5 per Congress). This represents a dramatic increase of four times more resolutions and recommendations adopted per Congress. Table 4.1 shows some examples of increase per region.

This increase is very similar to that of the global IUCN resolutions and recommendations adopted at each Congress, which has also multiplied by four. In other words, after the generalization of ICTs, IUCN Members adopted four times more resolutions and recommendations (per region and globally).

The process that leads to the creation of a motion entails much coordination, communication, and information sharing. As mentioned previously, each motion is the result of the work of at least six IUCN Members, which agree on a common text. Given the geographical spread of IUCN Members, this coordination is possible only by using affordable means of communication. The process of proposing a motion through traditional forms of communication such as onsite meetings, normal postal services, and intercontinental telephone is too costly and accessible only to Members with large budgets.

The generalization of new ICTs improves the capacity for coordination, communication, and information sharing. This led IUCN Members to submit a larger number of motions to be considered and adopted on site. Therefore it can be concluded that the increase in the number of resolutions and recommendations can be explained by the generalization of new ICTs among IUCN Members and the IUCN Secretariat since 2000. Since ICTs supported Members to submit more

Table 4.1 Evolution of the average number of resolutions and recommendations about regions adopted per Congress before and after 2000 (until 2012 Congress)

Topic of resolution and recommendation: regions	Average number of resolutions and recommendations adopted per Congress before 2000	Average number of resolutions and recommendations adopted per Congress after 2000
Africa	1.85	7.5
South and Meso America	2.1	14.25
South and East Asia	1.55	10
West Asia	0.4	8.7
West Europe	1.85	8.75

motions to be adopted, they also contributed to an increased legitimacy of this global environmental governance mechanism.

4. Online discussion platform

In 2016, the organization decided to digitalize completely the submission,[41] discussion, and voting parts of the motion process for several reasons. First, the number of motions was increasing with each Congress and required more and more time on site. The organization came to realize that agreement could in fact be quickly reached for some motions and that this could easily take place prior to Congress. This would not only reduce the number of motions to be discussed on site, but allow more time for IUCN constituents to address other issues. In addition, ICTs allow information to be shared transparently and simultaneously throughout the planet; this is particularly crucial for an organization with its constituents spread over the globe. Furthermore, using digital technologies is cost effective for the organization and for its constituents, in particular IUCN Members. Sharing information and enabling constituents to debate and vote online require fewer resources for an organization than doing so through traditional means such as onsite meetings and postal services. It also reduces the carbon footprint of the Union as a whole. For its Members, to be able to access information, coordinate with others, propose new motions, debate, and vote through digital technologies is much more affordable. This is particularly important for IUCN, since more than 1,100 NGOs are Members of the Union, and many of them are small NGOs with limited resources. In other words, the digitalization of the motions process allows it to be more accessible to all.

The new digitalized version of the motions process was implemented prior to the 2016 World Conservation Congress. It includes two parts: discussion and voting processes. All motions accepted by the Motions Working Group are submitted to the online discussion forum, but some will also be debated at the Congress as explained further.

The online discussion system is a web-based platform[42] where all IUCN constituents can express their concerns, debate pros and cons, and suggest specific changes for each motion. All constituents can transparently see who supports, comments on, and edits motions. For each amendment, the category of Member is indicated (A, B, or C). The online discussions are facilitated by individuals chosen by the Motions Working Group. Each motion discussion is facilitated by an individual called a facilitator who ensures that discussions are transparent and fair, and lead to a text that can reach a consensus through the electronic voting system. The facilitator is chosen among IUCN staff and Commission Members who volunteer for these positions that exist only prior to and during each Congress. The Motions Working Group distributes the motions. Facilitators are assigned a number of motions based on their availability, their acquaintance with the topic, and the difficulty of the motion; each facilitator is in charge of only one potentially controversial motion or up to three more straightforward ones.

The main communication channel for facilitators is through the online discussion platform's announcement feature, which sits at the top of the webpage of each motion. Participants can view these announcements, which are a convenient way for the facilitator to inform Members of a revised version, address a specific issue, or publicize a Skype call about a specific aspect of the motion. The facilitator can also make comments in order to help reach a consensus.

Although IUCN Commission members, National Committees, and staff can take part in the debate, they can do so only in an advisory and support capacity. In other words, they can comment on any part of the motion and suggest text edits, but only to provide technical advice and support the discussion.[43] Only Members are allowed to make amendments and take part in the final vote.

Online discussions take place during two distinct periods, called readings. The first one focuses on the original version of the text as proposed by the sponsor and co-sponsors. In this reading, constituents can express their concerns and propose changes to the first version of the text. On the fifth week of the online discussion, each facilitator compiles all comments and changes that constituents agree on and incorporates them into the original text. This second version of the text is then submitted to the second phase of the discussion, called second reading, which starts on week six and focuses on the new edited text. Similarly, constituents can discuss and propose changes during this second reading. On week nine, the online discussion closes, and each facilitator incorporates the last edits before sending the final text to the Motions Working Group with a summary of the main points of discussion and the status of the debate. The Motions Working Group then decides which motions are suitable for an electronic vote and which motions should be sent to Congress for further debate.

If some constituents choose to debate a motion outside of the online discussion system, the facilitator must report their comments in the online discussion system so that everyone may transparently have access to the same information and can understand the rationale behind these edits.

On the online discussion platform, IUCN constituents must log in with their IUCN account. They have access to all motions, but they can also subscribe to those of particular interest in order to receive automated email notifications about the latest changes, comments, and support. For each motion, a space is dedicated to announcements from the facilitator, the status of the version discussed (first or second reading), the body of the motion with the explanatory memorandum if available, sponsors and co-sponsors, and lastly comments. On the online platform, the title, preamble, and operative paragraph(s) can be discussed and edited. The online discussion language (called working language) is defined by the facilitator and usually corresponds to the language of the motion. IUCN counts three official languages (English, French, and Spanish), and comments can be posted in all three. The working language remains the same for the two readings. Although it is possible to use Google translation tools to change the text of the motion, the official text is considered in its original language only. No Google-translated version may be considered official.

For a comment to be valid, the constituent must be identified (Member or other constituent), the comment must be identified as being a general remark or a specific text edit, and lastly it must be indicated whether the comment concerns the whole motion, the title, the preamble, or a section of the motion. Other participants can then filter the remarks and edits by type of constituent and part of the motion addressed by the comment. In the case of text edits, IUCN recommends copy-pasting the part of the motion that needs change into the comment section and then adding new text in bold and/or striking through parts of the text that must be deleted according to the commenter. All constituents then have the possibility to "like" the comment to indicate they support this change. If a constituent disagrees with the proposed edit, he/she must add a new comment with an alternative suggestion.[44]

On weeks five and nine, the facilitator integrates all comments (remarks and edits) into the text of the motion. He/she can organize side discussions about specific comments to help reach consensus, but each side discussion must be reported in the online discussion system. After week nine, the new text is sent to the Motions Working Group for consideration. In addition to the new text, the facilitator is required to prepare a succinct electronic report providing some background information on the motion's debate and how far from consensus the participants are. The report should address the following questions: how did the discussion go? Did Members quickly agree? What was the main issue of contention? Was there general agreement on the text at some point? Are Members against the core topic of the motion rather than only some parts of it? Is the motion ready for an electronic vote? This information will help the Motions Working Group decide whether the motion should be submitted to an electronic vote or if it should be sent to Congress for further discussion.

In terms of effective digital consultation, a minority of Members commented on the online discussion platform. This does not mean that only a minority of Members participated in the online voting process, but rather that a minority of Members made comments. The following results exclude comments from facilitators.

As shown in Table 4.2, thirty-six Members from category A commented at least once: 19 (out of 89 or 21%) states and 17 (out of 128 or 13%) governmental agencies made 979 comments. One hundred sixty-eight Members from category B commented at least once: 29 (out of 112 or 26%) international non-governmental

Table 4.2 Distribution of comments on the online discussion platform per Member category[45]

Members	Number of comments	Number of Members who commented	Total number of Members	% of Members who commented
Category A	979	36	217	16.6%
Category B	1572	168	1084	15.5%
Affiliates	9	2	48	4%
Total	**2560**	**202**	**1349**	**15%**

organizations and 139 (out of 972 or 14%) national non-governmental organizations made 1572 comments. Two (out of 48 or 4%) affiliates (which can comment but not vote) made nine comments in total.

This first series of results shows a large number of Members across all categories that commented during the online discussion. This is a first positive result in the sense that the online discussion platform did not favor or disfavor any one type of Member. On average, 15% of all Members made a comment online, which corresponds to the e-consultation aspect of the digital participation definition. In terms of geographic distribution, the results are similarly homogenous, with no major differences between regions. In any case, the geographic aspect is less relevant here since it is largely influenced by the topic of the motions: for instance, Members from West Europe will comment more if more motions about Europe are submitted. Moreover, the top ten Members that commented the most account for about 40% of all comments. Table 4.3 shows the top ten IUCN Members providing the most comments on the online discussion platform.

These results provide a good illustration of how unique IUCN's motions process is: states and non-state actors making propositions hand-in-hand for nature conservation. They also show that the process was not overwhelmingly dominated by a small

Table 4.3 Top ten Members providing the most comments on the online discussion platform[46]

Name of Member	Type of Member	Category of Member	Number of comments	Percentage of total comments
US Department of State, Bureau of Oceans and International Environmental and Scientific Affairs	State	A	283	11%
Parks Canada Agency – Agence Parcs Canada	State	A	109	4%
Ministerio de Relaciones Exteriores	State	A	107	4%
Environment and Conservation Organizations of New Zealand	National NGO	B	100	4%
Ministry of Foreign Affairs of Japan	State	A	88	3%
Fundación Biodiversidad	National NGO	B	85	3%
Eco Redd	National NGO	B	79	3%
Australian Government Department of the Environment	State	A	65	2.5%
Center for Environmental Legal Studies	National NGO	B	59	2%
Fédération des Associations de Chasse et Conservation de la Faune Sauvage de l'UE	International NGO	B	52	2%

number of Members. The top ten Members count five states (category A) and five non-state actors (category B). This is an important point to highlight, as it indicates that the organization succeeded in developing an online process that could benefit all Members regardless of their category. When discussing the positive impact of ICTs on participation, this first set of results shows that digital technologies, when based on well-designed processes, can offer an efficient alternative to onsite procedures.

Furthermore, 206 (out of 1349 or 15%) Members made 4,133 interventions on the online discussion platform as shown in Table 4.4. Interventions comprise comments, but also amendments, announcements, endorsement of comments ("likes"), and suggestions. This number is consistent with previous results and shows that overall 15% of all Members took part in the discussion. Concerning other participants, 10% of all interventions were made by facilitators and 10% by Council, Commissions, National Committees, and staff, which shows that the major part of the discussion (80%) was led by Members.

The rather limited number of interventions and comments (made by only 15% of IUCN Members) can be explained by a number of reasons. First, as it is the case in traditional settings, some Members take part in the process only by reading and following the debates but without intervening to make comments. Moreover, Members might not necessarily have felt the need to comment on the platform if they agreed on previous comments or with the text of the motion. Also, Members knew that if a consensus was not reached, the motion would be further discussed on site at Congress. The limited participation could also be explained by the fact that this new version of the motions process was implemented for the first time in 2016, and would therefore need to be used for future Congresses so that Members could adapt to the new procedure.

The online discussion system allowed more Members to contribute to the discussion than was the case with the previous onsite procedure. Indeed, in 2012, only 200 interventions by about 100 Members led to the adoption of a much larger number of motions.[48] This is in comparison to the 4,133 interventions from 206 Members in 2016. Although the percentage of e-consultation is low (15%), it corresponds in fact to a much wider number of IUCN Members that effectively took part in the discussion. In the past, there was a relatively limited number of Members that participated, but in actual fact ICTs did increase the levels of

Table 4.4 Distribution of interventions per participant on the online discussion platform[47]

Type of participant	Number of interventions	% of interventions
Facilitators	544	10%
Council members	38	1%
IUCN Members	4133	80%
Other participants: Commissions, National Committees, and staff	446	9%
Total	**5161**	**100%**

participation, both the number of interventions and the number of Members, with 20 times more interventions and twice as many Members participating.

5. Online vote on motions

When the discussion period is over, the Motions Working Group refers each motion either to the Members' Assembly (because of the topic or divergent opinions) or to an electronic vote. In the latter case, this means that Members have reached a consensus on the final text of the motion, or alternatively that the text of the motion did not trigger any amendments. If there are amendments, it means that the text has not reached a consensus among Members. Further to the electronic discussion, the Motions Working Group decides on a limited number of motions to be discussed and voted on at Congress. The selection is based on three criteria: (1) the importance of the issue for conservation warrants a debate at Congress, (2) a high number of divergent proposed amendments did not lead to a consensus through the electronic voting system, and (3) the motion concerns IUCN governance and the IUCN Statutes and Rules of Procedure.[49] For the 2016 IUCN Congress, 99 motions[50] were accepted by the Motions Working Group and submitted to the online discussion forum; of these, 85 were submitted to an electronic vote.

The electronic vote must be opened and closed prior to the Congress dates.[51] Once adopted electronically by Members of category A and Members of categories B and C combined, the motions have the same validity as the ones adopted through the traditional onsite process. Through a vote at Congress, the Members' Assembly will record "en bloc" all motions voted on electronically, formally making them effective on that date.[52]

At Congress, if motions need further clarification and discussion, the Resolutions Committee can decide to refer a motion to a specific committee or ad hoc contact group for its review and advice. The new text emanating from such a group will be transmitted to the Resolutions Committee prior to its presentation to the Members' Assembly, which will subsequently vote on the new text.[53]

For the motions that are submitted to an electronic vote, eligible IUCN Members, also called "authorized vote holders,"[54] receive by email a unique URL, through which they can access the electronic ballot and cast a vote for their organization or institution. For some motions, a single question is asked: do you adopt the motion? The vote holder can respond by answering yes, no, or abstain. If a Member does not vote, it is considered an abstention. For other motions, multiple questions are asked. This is due to the existence of amendments to a motion. Amendments are alternative text options that the Motions Working Group decides to submit separately for two reasons: either they were suggested at the last stage of the electronic discussion and the available time did not allow for proper debate; or there was a limited prospect of reaching a better consensus on site among the proponents of the several text alternatives.

Each electronic ballot is dedicated to a specific eligible IUCN Member. The organization does not allow voting by proxy, which implies that the unique URL should not be shared with others. Each ballot shows a list of all motions and amendments that require voting. For each motion and amendment, the vote holder must choose between the three vote options mentioned previously, save the choice, and move to the next one. Voters have the possibility, at the time of voting, to access the text of the motion once more through a dedicated link. At the end of the whole process, Members must submit their votes. The electronic voting system allows for a final review of all votes, as well as the possibility of making some final changes, before he/she can confirm and submit his/her votes, which are then recorded appropriately. Once this is completed, there is no further opportunity to modify the votes. The Member then receives an email confirming all votes.

Although motions with amendments represent a minority, their adoption procedure is specific and more complex than for other motions. In 2016, there were only 12 motions with amendments that were submitted to an electronic vote. The procedure has some similarities with the onsite procedure (when amendments are considered by the Members' Assembly during an IUCN Congress), but also some differences due to the nature of the online voting system, as discussed below.

Traditionally, meaning on site, voting on amendments takes place in several rounds. Members vote on amendments first, and then on the motion with the amendments if consensus was reached on a specific text that includes some or all of the amendments. Amendments can address different parts of the motion, or alternatively, several amendments can tackle the same part of the motion. This latter case requires a specific voting process: Members first vote on the amendment that differs the most from the text. If that amendment is accepted, the other amendments concerning this part of the text are disregarded and the voting continues with any amendments that tackle other parts of the text. If the amendment is rejected, then Members vote on the second amendment relating to the same section of text. The process continues until there are no more amendments for this specific text. If no consensus is reached, the motion is rejected. If a consensus about this part of the text is reached, then voting continues with other amendments, if any, about other parts of the text. At the end, a vote is taken on the new amended version of the text, which will become a resolution (for IUCN's constituents to act upon) or a recommendation (for the consideration of other stakeholders).

This process is possible on site but impossible online since the various results of the voting for each individual Member are not known simultaneously for all Members, given that they might not vote at the same time. Members have two weeks to vote, and therefore the final vote on each amendment is not known until the end of the voting period. For this reason, IUCN created a new procedure specifically dedicated to online voting on amendments. This procedure consists of a series of "what if" questions that Members must respond to with a yes, no, or abstain. These hypothetical questions allow the electronic voting system to take into consideration all possible voting options.

Many aspects of the online vote differ from the onsite voting procedure described above. First, each vote is not merely on an amendment, but on the motion with the amendment. Members must answer a question such as: "Do you adopt Motion 001 with amendment 1a?" The next question will be: "Do you adopt Motion 001 with amendment 1b?" and so on. Amendments are presented in the same order as the text they address. When several amendments tackle the same part of the text, the version most different from the original text is first proposed for a vote, similar to the onsite voting procedure. This sequencing is also important for the online voting procedure because it is based on a probability "what if" question system: in other words, if Members reach a consensus on the first question (Motion 001 with amendment 1a), the following question (Motion 001 with amendment 1b) will not be considered. It is therefore crucial that the same rule applies on site and online, to allow an equivalent treatment of all motions and amendments, as stated by the IUCN Statutes and Regulations (articles 60 and 61).

Members must answer all questions concerning all amendments. As discussed previously, since voting takes place over a period of time, this is the only way the organization can integrate Members' choices. This differs from the onsite process in two ways. First, if a consensus is reached on an amendment, other amendments for the same part of the text are disregarded on site. Online, since Members vote on a motion with a specific amendment, they must cast their vote on all possible versions of a motion with amendments. If a vote holder chooses to adopt Motion 001 with amendment 1a, he/she will still have to vote on Motion 001 with amendment 1b, 1c, etc., including the amendments that contradict amendment 1a that he/she adopted. This could be perceived as time consuming, but it is the best way to ensure that all options are considered. Lastly, after considering all amendments, Members are asked to vote on the motion without any amendments. This provides the possibility to adopt the original text or to reject all amendments and the original text.

Once all votes are consolidated, the organization publishes the results online. The final version of the motion consists of the original text modified by all amendments adopted; amendments not adopted will not be part of the final version of the text. In terms of number of votes and percentages, IUCN publishes the results for motions and amendments that were adopted. The organization also shows results for amendments that were not adopted, but only for those that preceded the amendments that were adopted. This means for instance that if three amendments tackled the same part of the text, and the second one was adopted, IUCN will show the voting results of amendment 1a (rejected) and 1b (adopted) but not 1c. Results of this motion without any amendments will not be published either.

Members have the possibility to add a written statement along with their vote. At the end of each electronic ballot, a dedicated space allows them to explain their voting choices. They can also send an email to IUCN staff before the closing date of the voting period. Their explanation will then be recorded and published with the results of the electronic votes. In addition, IUCN publishes the results of all individual votes: all IUCN constituents can therefore see who voted in favor and who voted against each motion and amendment. The results are published on the IUCN intranet, which gives access to all IUCN constituents.

The online voting period for 2016 was from 3 August at 12 noon GMT/UTC until 17 August at 12 noon GMT/UTC, and a total of 85 motions were adopted. Seven hundred eleven Members voted; this shows that while a moderate 15% of all Members took part in the discussion process, more than half of all Members

(52%) voted, accounting for 68.6% of voted representation.[55] This latter percentage takes into consideration the affiliates that cannot vote and the distribution of votes between states and their governmental agencies, as described previously. Furthermore, all eligible Members confirmed and adopted these motions "en bloc" on site at Congress. This mitigates the limited number of comments and interventions and reinforces the success of the online discussion platform.

Further to the online discussion, 14 motions out of 99[56] were sent to Congress:[57] six motions for warranting debate at the global level and eight motions that did not reach consensus. This leads to the conclusion that the online discussion platform helped not only to increase the level of transparency and participation, but also to reach a consensus for more than 91% of all motions. With this level of consensus, the online voting process developed by IUCN confirms that ICTs can enable a global organization to increase the transparency and legitimacy of its decision-making processes. Multi-stakeholder processes such as IUCN's governance model that include state and non-state actors are often considered slow, costly, and ineffective in the sense that they rarely lead to a consensus. This analysis clearly shows that ICTs can mitigate the traditional disadvantages of a multi-stakeholder process and enable a decision-making process that benefits fully from the additional transparency, inclusiveness, and representation of multi-stakeholder governance models.

6. Concluding remarks

Legitimacy needs accountability, access to information, and representative participation. As discussed in this chapter, IUCN's progressive digitalization of some of its governance mechanisms, namely the motions process, provides increased legitimacy to the decisions taken by the organization. Given its transparency, inclusiveness, and representation, IUCN's motions process is a unique multi-stakeholder forum, where a large part of the environmental community meets to discuss sustainability and nature conservation. It includes some of the major governmental and non-governmental environmental actors, including states, governmental agencies, international and national NGOs, but also environmental experts from IUCN Commissions and National Committees. Although only IUCN Members can cast votes, all these constituents can participate in the discussion that leads to the vote, which contributes to enriching the debates and reinforcing the transparency of the process.

The online discussion platform provides a unique digital space where part of the environmental community can debate and vote on proposals about nature and biodiversity conservation. Other international arenas exist such as UN conferences or CBD COP meetings. However, none of them recognizes civil society and states as formal members with equal voting rights. And none of them has developed such a transparent and inclusive online discussion and voting platform that allows its members to reach a consensus on more than 90% of the topics debated.

If this case of digital participation is conclusive, it is also due to two factors associated with technology adoption. Indeed, participating in the electronic discussion system required Members (and IUCN constituents) to adopt a new technology. This is not as straightforward as it may seem. Indeed, a barrier often prevents users from adopting a new technology, whether this is due to its complexity or to the user's limited understanding, competence, or time. This barrier can be either psychological (it seems complex and difficult) or technological (it requires a lot of knowledge and experience to use it). For both cases, the electronic discussion system was successful in developing a technology that was accessible to all constituents and could be easily used.

Furthermore, adopting a new technology, overcoming the barriers mentioned previously, and changing the habit of discussing and debating motions on site require clear and direct motivation. Indeed, no one will change a habit if there is no motivation to do so. This means that in the case of the electronic discussion system, Members had to clearly see the direct benefits of adopting this new technology. One could argue that the decision was made by Members in 2012 during the Congress in Jeju, Republic of South Korea. However, the levels of participation and the consensus found online would indicate that Members could clearly see the benefit of adhering to/taking advantage of/taking part in this new digital participation process.

The digitalization of the motions process, from submission to voting, also sheds light on the impact of ICTs on the participation of state and non-state actors in a multi-stakeholder governance process. Thanks to digital technologies, CSOs, which traditionally have fewer resources than states, can gain additional visibility and take a stronger role in these governance processes, which supports SDG Target 16.6, "Develop effective, accountable and transparent institutions at all levels"[58] and SDG Target 16.7, "Ensure responsive, inclusive, participatory and representative decision-making at all levels." Their use of ICTs also allows them to better cooperate with other actors and help achieve SDG Goal 17 entitled "Revitalize the global partnership for sustainable development."[59]

Since ICTs can help IUCN Members to reach a consensus in a more efficient way, it is yet to be determined if they lead to better implementation of the decisions taken. Although it is well known that inclusiveness of a governance process triggers better ownership of the decision and increased legitimacy, more research should be conducted in the case of the IUCN resolution and recommendation process to verify this point.

Furthermore, the digitalization of the motions process implies that all Members of the organization have equal technological capacity, in terms of internet access but also education and knowledge. This is yet to be demonstrated. Indeed, the digital divide shows that all parts of the world are not equally connected to the internet. This digital divide also applies to the difference of connectivity and technological knowledge between cities and countryside and between age generations. In this context, the organization will need to make sure that the digitalization of the motions process offers equal opportunities to submit, discuss, and vote to all IUCN Members. This can be done through either online training or help-desk capacity.

Lastly, the digitalization of the motions process raises questions about the role of CSOs in global environmental governance processes. If civil society organizations gain additional capacity to take part in global governance mechanisms, they challenge the role of the state. States have the legitimacy to make decisions on the international stage since they represent their citizens. However, it can be argued that some states also represent sometimes the specific interests of some parts of their populations, such as the business sector. Moreover, if CSOs have the legitimacy to participate as scientific authority, field expert, watchdog, influencer, and advocate, they still represent specific interests. They can defend a cause such as marine biodiversity or fight to combat climate change. But they still represent the interests of a specific group of citizens. In that context, a diversity of opinion, and therefore a multi-stakeholder approach, is the best guarantee to avoid that specific interests take over the governance of public goods such as the environment.

Notes

1 Holdgate, M., 1999. *The green web: A union for world conservation.* London, UK: Earthscan Publications, p. 24.
2 IUCN, 2018. *Rules of procedure of the world conservation congress*, Part. 7, Paragraph. 66, p. 21. www.iucn.org/sites/dev/files/iucn_statutes_and_regulations_january_2018_final-master_file.pdf
3 IUCN, 2016. *IUCN world conservation congress.* https://2016congress.iucn.org/programme/forum.html [Accessed 17 June 2018].
4 IUCN, 2018. *Rules of procedure of the world conservation congress*, Annex, Part. 2, Paragraph. 2, p. 32. www.iucn.org/sites/dev/files/iucn_statutes_and_regulations_january_2018_final-master_file.pdf
5 Ibid, Part. 5, Paragraph. 20, p. 10.
6 IUCN, 2018. *The impact of IUCN resolutions on international efforts*, pp. 11–17. https://portals.iucn.org/library/sites/library/files/documents/2018-011-En.pdf
7 World Conservation Congress, 2012. *Resolutions and recommendations adopted.* Gland, Switzerland: IUCN.
8 SDG Target 16.6. *Develop effective, accountable and transparent institutions at all levels.* See more here https://unstats.un.org/sdgs/indicators/indicators-list/
9 Xprize. *Op Cit.*
10 Cambridge Dictionary online. https://dictionary.cambridge.org/dictionary/english/inclusiveness
11 For a more detailed description of IUCN Members, please see here. www.iucn.org/about/union/members/how-become-member-iucn
12 IUCN, 2018. *The impact of IUCN resolutions on international efforts. Op Cit.* p. 10. https://portals.iucn.org/library/sites/library/files/documents/2018-011-En.pdf
13 IUCN, 2018. *Rules of procedure of the world conservation congress*, Annex, Part. 7, Paragraph. 48, p. 39. www.iucn.org/sites/dev/files/iucn_statutes_and_regulations_january_2018_final-master_file.pdf
14 Ibid.
15 Affiliate Members are government agencies and national and international non-governmental organizations without voting rights.
16 However, the explanatory memorandum shall not form part of the motion or be put to the vote according to: IUCN, 2018. *Rules of procedure of the world conservation congress*, Annex, Part. 7, Paragraph. 40, p. 50. www.iucn.org/sites/dev/files/iucn_statutes_and_regulations_january_2018_final-master_file.pdf

17 IUCN, 2016. *How the world conservation congress motions process works*, p. 4. http://cmsdata.iucn.org/downloads/information_on_motions_process.pdf
18 The Motions Working Group becomes the Resolutions Committee at Congress.
19 IUCN, 2018. *Rules of procedure of the world conservation congress*, Annex, Part. 7, Paragraph. 52, p. 40. www.iucn.org/sites/dev/files/iucn_statutes_and_regulations_january_2018_final-master_file.pdf
20 Ibid, Paragraph. 54, p. 41.
21 IUCN, 2012. *Resolution WCC-2012-Res-001 entitled Strengthening the motions process and enhancing implementation of IUCN Resolutions*. https://portals.iucn.org/library/sites/library/files/resrecfiles/WCC_2012_RES_1_EN.pdf
22 IUCN, 2016. *WCC-2012-Res-001 – activity report*. https://portals.iucn.org/library/node/45227
23 IUCN, 2018. *Rules of procedure of the world conservation congress*, Annex, Part. 7, Paragraph. 2, p. 2. www.iucn.org/sites/dev/files/iucn_statutes_and_regulations_january_2018_final-master_file.pdf
24 Ibid, Paragraph. 54, p. 41.
25 Ibid, Paragraph. 13, p. 7.
26 Ibid, Paragraph. 54, p. 41.
27 Ibid, Part. 5, Paragraph. 30bis, p. 13.
28 IUCN, 2018. *IUCN statutes*, Part. 5, Paragraph. 34, p. 13. www.iucn.org/sites/dev/files/iucn_statutes_and_regulations_january_2018_final-master_file.pdf
29 IUCN, 2018. *Rules of procedure of the world conservation congress*, Part. 5, Paragraph. 35, p. 13. www.iucn.org/sites/dev/files/iucn_statutes_and_regulations_january_2018_final-master_file.pdf
30 Ibid, Paragraph. 35bis, p. 13.
31 As of 17 June 2018. www.iucn.org/about/union/members/who-are-our-members
32 Poate, D., Gregorowski, R., Blackshaw, U., 2011. *External review of IUCN 2011*, Final Report. London, UK: ITAD, p. 115.
33 IUCN, 2018. *IUCN resolution and recommendation platform*. https://portals.iucn.org/library/resrec/help [Accessed 17 June 2018].
34 For more information see www.iucnredlist.org
35 For more information see www.iucn.org/theme/protected-areas/about/protected-area-categories
36 IUCN, 2016. *Our mission*. www.iucn.org/about
37 IUCN, 1984. *Recommendation GA 16 RES 018*. https://portals.iucn.org/library/sites/library/files/resrecfiles/GA_16_RES_018_Cooperation_with_other_Organizations.pdf
38 IUCN, 1988. *Recommendation GA 17 REC 026*. https://portals.iucn.org/library/sites/library/files/resrecfiles/GA_17_REC_026_Report_of_the_World_Commission_on_En.pdf
39 For more information, see the various versions of IUCN Statutes and Regulations. www.iucn.org/about/world-conservation-congress/congress-archives
40 The regional scope "global" was selected when conducting the search on the IUCN resolution and recommendation online platform.
41 The submission of a motion is done online and this functionality was already in place previously.
42 IUCN Online Discussion Platform. https://portals.iucn.org/congress/assembly/motions. [Accessed 13 June 2018].
43 IUCN, 2016. *The 2016 world conservation congress motions process – facilitating the online discussion of motions*, Gland, Switzerland: IUCN, p. 2.
44 IUCN, 2016. *The 2016 world conservation congress' motions process – 3*. The Online Discussion, Gland, Switzerland: IUCN, pp. 1–8.

45 IUCN Online Discussion Platform. https://portals.iucn.org/congress/assembly/motions [Accessed 13 June 2018].
46 Ibid.
47 IUCN, 2016. *IUCN world conservation congress 2016 communication by the motions working group on the online discussion*, Gland, Switzerland: IUCN, p. 1.
48 Ibid.
49 IUCN, 2018. *Rules of procedure of the world conservation congress*, Annex, Part. 7, Paragraph. 45bis, p. 38. www.iucn.org/sites/dev/files/iucn_statutes_and_regulations_january_2018_final-master_file.pdf
50 These 99 motions did not include the motions that address IUCN Statutes and Rules of Procedures.
51 IUCN, 2018. *Rules of procedure of the world conservation congress*, Annex, Part. 7, Paragraph. 62quinto, p. 44. www.iucn.org/sites/dev/files/iucn_statutes_and_regulations_january_2018_final-master_file.pdf
52 Ibid.
53 Ibid, Paragraph. 56, p. 42.
54 IUCN, 2016. *The 2016 world conservation congress' motions process – 4: The electronic vote prior to congress*, Gland, Switzerland: IUCN, pp. 1–13.
55 IUCN, 2016. *IUCN world conservation congress 2016 communication by the motions working group in preparation of the congress*, Gland, Switzerland: IUCN, p. 1.
56 In addition to these 99 motions, seven motions were sent to Congress for statutory reasons: they addressed IUCN Statutes and Rules of Procedure.
57 IUCN, 2016. *IUCN world conservation congress 2016 communication by the motions working group on the online discussion*, Gland, Switzerland: IUCN, p. 2.
58 SDG Target 16.6. *Develop effective, accountable and transparent institutions at all levels*. See more here https://unstats.un.org/sdgs/indicators/indicators-list/
59 Xprize. *Op Cit.*

5 Blockchain and environmental civil society organizations

1. Introduction

At the convergence of the physical, digital, and biological worlds, the concept of the fourth industrial revolution aims to describe how technological innovations are transforming all aspects and actors of society. These changes are unprecedented in terms of speed (exponential growth), scope (disruptions to almost every industry in every country), breadth, and depth (transformation of entire systems of production, management, and governance).[1] Blockchain is one of the most promising technologies, with the potential to greatly enhance the role of civil society as service provider and knowledge broker in global environmental governance.

Blockchain[2] is a global digital ledger that stores any type of transaction between two entities and individuals in a certifiable and enduring way. In other words, it is a growing list (a chain) of records (blocks)[3] and "the first native digital medium for value, just as the internet was the first native digital medium for information."[4] First presented in a paper in 2008 by Satoshi Nakamoto,[5] blockchain is the underlying technology that supports the creation and distribution of the cryptocurrency bitcoin. However, it is potentially much more than this: this globalized and distributed database can record any type of information and value – money, land ownership titles, transactions from a food supply chain, certification of origin for plants and species, scientific discoveries, patents, and even votes – on millions of devices throughout the planet.

Although we are in the early stages of experimentation and adoption, this technology is often compared with the early developments of the internet. Indeed, the success of internet applications such as Amazon, Facebook, Snapchat, and Netflix, to name only a few, was not foreseen in the early years of internet development. Nevertheless, thanks to the internet and its later applications, vast segments of society and the economy were changed.

Similarly to the internet, websites, and social media platforms, which supported the role and participation of civil society on the international stage,[6] blockchain technology promises to further reinforce the capacity of the nonprofit sector. According to the UN Climate Change secretariat (UNFCCC), blockchain technology can contribute to enhanced climate action and sustainability, in particular in terms of trust, transparency, incentive, and financing.[7]

According to William J. Sutherland et al.,[8] blockchain technology is expected to have an impact on some aspects of global environmental governance. Indeed, this team of experts from multiple scientific and policy backgrounds conducts a yearly horizon scan of global conservation issues, and identifies emerging topics that are likely to affect global biological diversity, the environment, and conservation efforts in the medium and long term. In 2017, they identified blockchain technology as a technology that can improve renewable energy production and distribution, transparency of endangered species trade, and field project funding.[9]

This chapter will first briefly describe blockchain technology, and then it will focus on three blockchain applications in global environmental governance. Indeed, this technology allows civil society to develop promising solutions to finance more efficiently the protection of the environment, to monitor more effectively food security and agreement implementation, and to nudge more effectively individuals toward more sustainable behaviors. Although an exhaustive list of all applications is difficult to produce, this chapter aims to illustrate how dynamic civil society is today. This chapter will also relate some of its applications to the Sustainable Development Goals (SDGs), showing how this technology can help achieve some of these goals and associated targets.

2. Blockchain in a few words

The high level of trust offered by blockchain platforms comes from its origin: it was first designed to support cryptocurrency exchange. Its objective was to avoid the *double-spend problem*: if one person sends money to another person or buys something online, the money must leave the buyer's account and go to the seller's account, otherwise the money might be spent twice, and the system becomes fraudulent. Before blockchain, third-party intermediaries solved this problem by clearing every transaction through one central database. Intermediaries include money transfer services (Western Union), commercial banks, and online payment platforms (PayPal for instance). These intermediaries invest a lot of time and demand substantive fees to perform the clearing and effectively transfer the funds.

Blockchain technology[10] functions differently: this distributed system consists of a community of agents[11] that store data and act as intermediaries of trust. It can be represented by a growing number of blocks of encrypted information linked to each other. This network of agents can be either public (open to everyone) or private (restricted to a certain number of pre-selected agents). Well-known open systems such as bitcoin and ethereum are publicly available for use. Others such as hyperledger fabric and multichain are close-ended.[12]

Satoshi Nakamoto designed a security system that allows each platform to become guardian of trust. Indeed, to add a new piece of information, a complex process must take place. First, each new block must refer to all the preceding blocks, making it impossible to omit or change previous data. A blockchain platform keeps track of all past transactions. Similar to the role of a public institution, it certifies the identity of the current bitcoin owner based on the history of all previous transactions.

Second, each new piece of information added to the blockchain must first go through a security process entitled *consensus mechanism*, which varies greatly from one platform to another, depending on its degree of openness. For instance, on an open blockchain platform such as bitcoin, the number of agents is expected to be quite large, anonymous, and untrusted. Therefore, the level of security must be very high. In this case, agents entitled to verify the new piece of information and to add it to the blockchain are randomly selected through a mathematical competition that takes place globally and requires large computing capacity. This avoids any collusion between the agents of the transaction (or the owners of the value) and the agents who verify and add it to the blockchain (who are called *miners*). The mathematical competition is called *Proof of Work* in the case of a bitcoin blockchain.

Satoshi Nakamoto's *Proof of Work* requires a lot of computing power for users to participate in the development of the platform, avoiding spams and denial-of-service attacks. It is based on Adam Back's method that requires email accounts to show proof of work when sending their messages: if the email is important, it must prove that the sender expended a lot of energy to produce it, which dramatically increases the cost of sending spams and other malwares. Other types of blockchain platforms can use other *consensus mechanisms* such as *Proof of Stake* (miners must invest in and keep some store value), *Unique Node List* (only entrusted agents who are pre-selected can verify and add new information to the blockchain), *Proof of Capacity* (miners must allot a specific volume of their hard drive to mining), and *Proof of Storage* (miners must share disk space in a distributed cloud).[13] In any case, the security process through which new information must go is highly elaborate and adapted to the degree of openness of the network, thereby encoding trust into the system itself.

Third, the members of this network, which are distributed throughout the planet, can download a copy of all content in the blockchain, leading to hundreds (or tens of thousands) of replicas of the same information, and therefore avoiding that one actor – a multinational corporation, or a state for instance – takes control of the whole system. It also makes it impossible to have a single point of failure as could happen with a centralized system.

Fourth, each transaction is time-stamped, recorded, and then stored. This means that all content is permanent, chronologically ordered, and available to all agents of the network.[14] Fifth, all content is transparent and visible to all agents. Everyone can see all transactions and information added to the blockchain. Each one of them sees all that the database contains as well as its history. However, transparency is combined with privacy, since members can remain anonymous by using a two-key cryptography system. The public key serves to identify the origin and destination of the transaction. It can be shared with anyone and can be published on a website or a blog. This is how agents are anonymously identified in the blockchain. The private key is confidential, and only the agent himself or herself has access to it.[15] The private key serves to unlock the transaction and deliver the value to the destination. Each transaction must be signed off with the identification of the seller.[16] It should be noted that a blockchain platform stores transactions, not balances.[17]

Since content in a blockchain is duplicated many times, there is no need for third-party authentication and control, and therefore no need for the centralization of authority. The platform itself is the new shared, authentic, and distributed authority that validates the veracity of information and values. In other words, governments could not arbitrarily freeze bank accounts, change property rights, or seize funds from political activists. Multinational corporations could not hide financial transactions in tax havens. Individuals could manage their own data individually (and ask for financial compensation).

This brief technical explanation will allow the next sections of this chapter to explore specific applications of blockchain technology in global environmental governance.

3. Financing environmental projects

Blockchain can help leverage the role of local grassroot actors in the global environmental governance architecture by considerably increasing the capital efficiency of donations to the green economy in developing countries. Indeed, one of the key components of blockchain technology is its capacity to store and keep track of all past transactions, creating a historical and immutable audit record that ensures high levels of trust among large numbers of agents.[18] This element of trust and transparency is crucial when it comes to financing field projects.

The importance of finance for the protection of the planet is well recognized. In 2016, UN Environment released a report on the implications of the Fintech industry for sustainable development,[19] focusing on three emerging technologies that could potentially affect global environmental governance: blockchain, the Internet of Things (IoT), and artificial intelligence (AI). Although this report was not the first one to stress the importance of green finance,[20] it is however the first one to acknowledge the disruptive role of digital technologies to finance the protection of the environment.

In a globalized world, trust is essential to ensure safe communications and transactions.[21] Many financial scandals have also led consumers to doubt the veracity of the information provided by their banks. Some financial institutions have developed socially responsible investing (SRI) strategies that respect certain ethical and sustainable rules, as part of the Environmental, Social, and Governance (ESG) movement. A plethora of labels exist to guide the investments of individuals and institutions, such as the Dow Jones Sustainability Index (DJSI),[22] United Nations Principles for Responsible Investment (UN PRI),[23] and Global Reporting Initiative (GRI) standards, used by more than 1,200 organizations for reporting on social, environmental, and economic performance.[24] Some of these labels and indexes tend to focus more on the sustainability quality of the financial product itself (aiming to incite the receiver of the investment to develop sustainable practices) than on informing the consumer, which makes some of these indexes hard to understand for the end client. In addition, the large number of existing labels increases their complexity and reduces transparency, hence their effectiveness, since the consumer is left to trust a third-party intermediary to read these labels and invest.

The current legitimacy crisis of some global environmental institutions is due in part to the lack of trust and the long distance between global decisional actors and the stakeholders most affected by their decisions, namely individuals and civil society. This distance, psychological and physical, tends to reduce local ownership and policy efficacy of global processes and decisions. So far, traditional intermediaries such as international organizations, banks, and governments have ensured this level of trust. Blockchain can do just the same, offering a high level of trust and security but through a decentralized solution.[25] Funds can be sent to the main beneficiaries directly, transparently, and with high levels of trust. Moreover, blockchain better connects donors with the field project they support. Although the transactions on a blockchain are anonymous, the transparency, combined with the possibility to use smart contracts, enables donors to feel closer to the field than they have ever been. Hence, the generalization of blockchain platforms will "allow us to put a nature fund in every pocket."[26]

Blockchain technologies can indeed help reduce resources spent on monitoring and tracking fraud, hence focusing on field projects and knowledge production. In 2011, former UN Secretary-General Ban Ki-moon argued that "corruption prevented 30 percent of all development assistance from reaching its final destination."[27] Although a share of this misuse occurs in developed countries, a significant portion is lost in the global south, where it is most needed. Fraud results from various causes, among which are inadequate government spending, corruption, and inefficiency at the time of delivering the products or services, which can lead to outbreaks of diseases in some countries and eradication of ecosystems and species. Tracking fraud is very costly and cannot apply to ongoing and future operations. It can be done only for past activities, which reduces its impact (but not its necessity). Blockchain platforms can dramatically change this situation, by enabling the efficient tracking of money, products, and services in real time. Thanks to this technology, international organizations can reduce fraud, save resources dedicated to monitoring and sanctioning fraud, and increase their positive impact: in other words, focus on their main mission, which is protecting the planet.

Individual and institutional donors from developed countries can directly support, at a low cost and without intermediaries, numerous individual and community projects in the global south. In the case of the two billion unbanked individuals in the world, the use of cryptocurrency allows donors to overcome this obstacle and fund very small restoration projects and local operations. This corresponds to SDG Target 1.4, "(. . .) appropriate new technology and financial services, including microfinance (. . .),"[28] and SDG Target 8.10, "(. . .) access to banking, insurance and financial services for all (. . .)."[29] Moreover, blockchain can help increase capital efficiency of remittances by removing the need for costly intermediaries, helping to reach SDG Target 10.c, "reduce to less than 3 per cent the transaction costs of migrant remittances."[30]

Blockchain technology enables not only the funding of local projects, but also funding for stakeholders who would not necessarily have engaged in a conservation project otherwise. In other words, anyone can become a conservationist today to restore urban ecology or prevent conflicts between humans and wildlife.[31] This corresponds to SDG Target 13.b, "(. . .) mechanisms for raising capacity for effective climate change-related planning and management (. . .),"[32] and 15.1, "(. . .) ensure the conservation, restoration and sustainable use of terrestrial and inland freshwater ecosystems (. . .),"[33] as well as SDG Target 15.c, "(. . .) increasing the capacity of local communities to pursue sustainable livelihood opportunities."[34]

The BitGive foundation aims to improve public health and protect the environment globally thanks to the use of bitcoin and blockchain technology.[35] Its main objective is to finance projects in developing countries. GiveTrack, its innovative donation platform, provides transparency and accountability to donors through real-time information about fund allocation. BitHope on the other hand is a Bulgarian civil society organization that generates funds for the nonprofit campaigns hosted on its website.[36] These platforms use smart contracts to ensure high levels of trust and deliver funds only when project delivery milestones are met.

The Natural Capital Finance Alliance (NCFA)[37] brings together UN Environment, Global Canopy, Fundacao Gerulio Vargas, and the finance sector to support the integration of natural capital considerations into financial decision making through driving innovation. Their objective is also to develop new instruments for the financial industry and better understand risks and find opportunities in the green economy.[38] NCFA intends to use a bitcoin blockchain platform to increase investments in projects to protect ecosystems and species biodiversity including rainforests, mangroves, and coral reefs. An Initial Coin Offering (ICO) will allow NCFA to raise capital in the form of digital currency. Each donor who acquires these digital coins can choose to identify the coins with a color that represents the biodiversity resources he or she wishes to protect. The donors can also choose the terms and conditions for their digital coins to be exchanged for cash: this could be deadlines or targets that must be reached. These coins representing biodiversity assets allow NCFA to digitize and monetize natural capital.[39]

Another example is the Forest Carbon Partnership Facility (FCPF), an alliance of governments, businesses, civil society, and indigenous peoples that aims at protecting forests throughout the world and decreasing emissions from deforestation and forest degradation.[40] In the global south, countries and various stakeholders are incentivized to protect their forests by receiving payments when reaching preselected targets. The meta-objective of this mechanism is to mitigate the negative impact of deforestation on global climate change. The Carbon Fund, managed by FCPF, supports financially various stakeholders such as forest-dependent indigenous peoples, other forest inhabitants, or the private sector, when emissions are reduced and forests are managed sustainably. In this context, blockchain technology reveals itself to be useful: digital coins can be issued to increase the capital efficiency of the incentives. Infinite Earth is a pioneering issuer of such digital coins, entitled Rimba Raya Biodiversity Reserve REDD+[41] credits. They allow the incentive to support the stakeholders in the field who are in direct contact with forests and whose actions have a very tangible impact on forest preservation. They also disrupt institutional carbon brokers, who profit from the lack of transparency and liquidity of the carbon market.[42]

The International Union for Conservation of Nature (IUCN), presented in depth in other chapters, has developed the IUCN Green List of Protected and Conserved Areas (the IUCN Green List) to encourage, achieve, and promote effective, equitable, and successful protected areas. To become part of this list, protected areas must abide by a set of components, criteria, and indicators, which provides an international benchmark for nature conservation.[43] Financing protected areas is also one of the objectives of the IUCN Green List, which aims to become a recognized conservation investment instrument to bring new resources to listed protected areas. The IUCN Green List Standard Token (GLS) is a blockchain-based smart contract platform developed to implement the IUCN Green List.[44] Token Generating Events (TGEs) will allow institutional and private donors to acquire GLS, which will subsequently enable IUCN Members, including civil society, to overcome the bootstrapping phase and leverage higher investments. Donors will have the capacity to easily monitor how their resources are used and how protected areas are managed. Smart contracts will indeed allow absolute trust about the execution of standards and criteria implementation stages.[45]

In many states, centralized producers and distributors tend to block local initiatives to produce green energy by keeping the price of green-produced electricity low in order to avoid further investments. EcoChain[46] is a platform for people to invest in all sizes of renewable energy projects (from utility-sized projects to social-good renewable energy projects). This green investment hub selects projects that are pre-vetted and open for investment. Each user can choose the level of investment, which is converted into bitcoin. Protective smart contracts then release funds only when specific targets are met by the project. The use of smart contracts ensures that the overall investment is secure. Smart contracts indeed have many advantages: they execute with or without the will, consent, or actions of the parties involved, and since they are automatic, it is not possible to withdraw from them.[47] These examples correspond to SDG Target 7.2, "(. . .) increase substantially the share of renewable energy in the global energy mix (. . .)."[48]

Thanks to blockchain, and the levels of trust and transparency it allows, donations are not only likely to increase but also transform the global environmental governance architecture, since traditional intermediaries who pool resources and then coordinate their implementation will not be necessary anymore. Donations can be sent directly and safely to local individuals and communities with no intermediaries. In that sense, a part of the administrative work done by international organizations is highly likely to be replaced by blockchain applications. Smart contracts enabled by blockchain platforms can automatically trigger transactions when certain conditions are met, such as delivering an item (that triggers payment) or passing the expiration date (that triggers a new shipment). Smart contracts allow international organizations and funding agencies to send only a small amount of money corresponding to each transaction. This reduces the administrative burden carried out by international organizations, and at the same time enables these organizations to avoid transferring large amounts of money to ministries and local public administrations that are more difficult to trace and control.

The following section examines how blockchain applications can greatly enhance the monitoring role of civil society organizations.

4. Effective monitoring

Blockchain technology supports civil society organizations in their monitoring roles, and in particular in overseeing food security, the toxic water supply chain, overfishing, forest management, the trade of endangered species, and the implementation of international agreements.

In a globalized world, products are often the result of the collaboration of many actors. Increasingly, consumers look for ethical and healthy products to buy. Traceability is key to certify their origin, production, and modes of transportation. The global market for ethical products including packaged foods, soft drinks, and hot drinks reached US$793.8 billion in 2015. Across all labels, sustainability labels are the most represented (three-quarters), and 97% of these refer to recycling. Clean labels certify that products are free of some harmful chemicals or ingredients, the most common of which are: artificial preservatives, colors, flavors, additives, and sweeteners; monosodium-glutamate (MSG); Bisphenol A; and Genetically Modified Organisms (GMOs).[49] Official organic certifications are also essential and already provide sound guidance for consumers to choose their products. However, controls on producers are sometimes rare and differ from one country to another.

Several food scandals have led consumers to doubt the quality of some products and the veracity of the information provided by producers and retailers. One agent of a global supply chain could compromise the quality of the final product without the knowledge of the retailer and, by extension, of the consumer. There is a need for transparency to allow the consumer to know the origin of the products, what type of agriculture and transformation processes were used (and their environmental footprint), and how employees were treated, to name but a few. These elements should not only determine the final price, but also help the consumer to make a choice. By allowing the consumer free and easy access to this information, and by ensuring the veracity of this information, trust can be rebuilt. As discussed in the preceding pages, blockchain can ensure the veracity of this information. But it can also ensure its availability to consumers, making global supply chains more transparent.

Blockchain technology allows every step of a product's life to be monitored and recorded, from raw material sourcing to final sale, providing individuals with reliable, transparent, and accurate information about their potential purchase. Each transaction is time-stamped and geo-localized. Each agent of the supply chain is identified and its actions recorded. This information is made available for everyone to see, and there is no way to corrupt it or change it. The inherent transparency of a blockchain platform can lead to greater accountability and better production and transportation practices. Hence, consumers can find out where a fish was caught, how far a tomato had to travel to be processed, how many transformations a packaged food went through, and if it really comes from a fair-trade producer. Each ingredient that is added (and its origin) is time-stamped and recorded. The end consumer can therefore find out where all the parts of a shoe originate, if they were made sustainably, and if natural components were used. The consumer can play an active role in shaping the future of global trade, and subsequently of our planet, through informed and transparent buying choices. This corresponds to SDG Target 12.8, "(. . .) people everywhere have the relevant information and awareness for sustainable development (. . .)."[50]

In this context, blockchain is of great use for civil society organizations that monitor food supply chains. Foodtrax[51] is a blockchain project that tracks food and develops high quality ethnic food concepts. Provenance[52] is another blockchain application that aims to build trust in goods and the supply chain, and help consumers choose products. For instance, it collaborated with the Indonesian fishing industry to trace sustainably caught fish. Based on the same principles of reinforcing trust and traceability, the start-up Everledger[53] certifies and monitors trade in diamonds, in order to decrease sales of stolen gems or conflict stones. Blockchain could also greatly simplify the work of customs officials since they could easily identify illegally traded animal parts or plants thanks to mobile DNA barcode scanners for instance.[54]

Blockchain technology can also monitor much more efficiently the supply chain of dangerous waste. For instance, when poisonous waste delivery is provided with a digital identity, each waste item can be tagged with a QR code and scanned throughout the transportation chain from the field to the recycling area. Once the product has reached its final destination, the transporter and the recycling manager (both part of the blockchain network) agree to finalize the transaction, which is added to the blockchain. From this moment onwards, there will be permanent proof of where the waste items transited, who transported them, when they arrived, and who accepted them. Each one of these individual transactions is added to the blockchain, which will make it impossible for these products to be lost, altered, or switched with imitations, without the network being informed. The blockchain platform can even contain crucial information such as loading and storing conditions. This corresponds to SDG Target 12.4, "(. . .) sound management of chemicals and all wastes throughout their life cycle (. . .)."[55]

Such a tracking and permanent recording system will also help to prevent out-of-stocks, essential in the case of emergency procedures and disaster risk management. Since every step of the supply chain is monitored and cannot be altered, blockchain platforms can help donors not only verify where their funds are allocated, but also pinpoint where there is insufficient staffing, or where dangerous items are inappropriately stored. This corresponds to SDG Target 12.3, "(. . .) reduce food losses along production and supply chains (. . .)."[56]

In terms of overfishing, the blockchain-based fish catch documentation system provides end consumers with complete transparency of the supply chain. Each fish is tagged electronically, allowing any stage of the transportation and trans-formation to be tracked and recorded: who caught the fish, where and when, who transported the fish, from where to where and when, who transformed the fish, etc. This accurate, reliable, transparent, and immutable record of the supply chain benefits not only the consumer, but also the local fishers, who can obtain credit loans based on the sustainable fish catch documentation and can receive digital money even if they do not have a bank account. Micro-payments, for each fish, are impossible within a traditional financial landscape, but entirely possible with a blockchain infrastructure. This allows local communities to avoid relying on costly intermediaries and receive a more honorable (and well-deserved) compensation for their work. In addition, the fish catch documentation can help governmental and international agencies to monitor and better control fishing quotas in some regions of the world.[57]

The Earth Twine-Stratis Platform is another blockchain platform dedicated to storing origin data and tracking the international fishing industry. Launched in 2017, it aims to reinforce the traceability of seafood products, the second largest commodity on earth,[58] targeting specifically unreported fishing mixed within the supply chain of legal products.[59] Similarly, the World Wide Fund for Nature (WWF) has applied blockchain technology to the tuna fishing industry in order to eliminate illegal catching and human rights abuses in the Pacific Islands region.[60] To avoid buying a tuna fish coming from unregulated fishing or from companies using slave labor, blockchain technology allows consumers to scan any tuna packaging with their smartphone to find out where and when the fish was caught, by which vessel, and using which fishing method. As WWF-Australia CEO Dermot O'Gorman stated: "Bait-to-plate transparency using the blockchain will mean there is no place to hide for illegal, unregulated and unreported fishing or those operators who use slave labour or impose horrific conditions."[61] This corresponds to SDG Target 14.4, "(. . .) effectively regulate harvesting and end overfishing, illegal, unreported and unregulated fishing and destructive fishing practices (. . .),"[62] and SDG Target 14.b, "(. . .) access for small-scale artisanal fishers to marine resources and markets (. . .)."[63]

Forest Stewardship Council (FSC) is a CSO that aims at promoting sustainable forest management, and whose members come from forestry companies, environmental groups, retailers, academia, and other civil society entities. This certification body allows consumers to choose wood stemming from responsibly managed forests.[64] However, the organization wishes to increase the transparency of its certification process and address misleading and false claims. "Millions of consumers and thousands of businesses trust the FSC label, and we must do what we can to ensure that we maintain that trust," says Kim Carstensen, Director General of FSC.[65] With this objective in mind, FSC is implementing a new strategy based on technology: tracking transactions between trading parties, developing an online claims platform for specific high-risk supply chains, and testing the viability of digital claims and blockchain technology.[66] This corresponds to SDG Target 15.2, "(. . .) implementation of sustainable management of all types of forests, halt deforestation, restore degraded forests (. . .),"[67] and SDG Target 15.3, "(. . .) restore degraded land and soil (. . .)."[68]

Blockchain can also support civil society organizations to better monitor the role of states when it comes to land rights. Governmental agencies keep track of all previous owners of a piece of land. This is one of the roles of public administrations, to verify and certify that a specific piece of land belongs to a specific person, who therefore has the right to sell it. The state verifies the ownership through the string of certificates that proves the traceability of the ownership of land, owner after owner. Blockchain can do just the same, offering a high level of trust and security but through a decentralized solution.[69] It allows civil society to develop innovative applications to attest to the existence and ownership of any type of value that can be digitalized, such as land claims registration, which is crucial in species-rich parts of the developing world where property ownership is frequently contested or where administrative institutions are not trusted.[70] This corresponds to SDG Target 1.4, "(. . .) ownership and control over land and other forms of property, inheritance, natural resources (. . .)."[71] In Ghana, the startup BenBen is developing land-title registries using blockchain technology. Georgia and Honduras are following the same path. Blockchain creates this persistent, authentic, and transparent digital identity, which can be shared with other applications and allow a better reward for biodiversity patents and rights to indigenous communities throughout the planet. This matches SDG Target 15.6, "(. . .) fair and equitable sharing of the benefits arising from the utilization of genetic resources (. . .)."[72]

Key to environmental treaties is their implementation. Once treaties are ratified and entered into force, member states or Parties to a Conference (COP) regularly meet to take stock of implementation advances and difficulties. In many cases, these discussions are based on voluntary reporting and monitoring reports from civil society organizations. In both cases, states could be inclined to manipulate data, avoid mentioning some elements in their reporting, or pressure civil society to cover for them by misreporting their lack of progress. Blockchain technology can help overcome these reporting issues. Since each transaction in the chain is time-stamped, data fraud and manipulation can be easily avoided. Monitoring reports from civil society and states' voluntary reporting can be added to the chain independently and transparently. Anyone can have access to these results and find out how the implementation of international commitments is advancing. In addition, once in the chain, data cannot be altered and exert additional pressure to declare realistic advances that will be checked many times in the near and more distant future. In this case, blockchain can significantly help implementation processes by increasing levels of transparency for reporting and monitoring processes. Since voluntary reports and external reviews can be permanently added to the blockchain, the incentive to not declare, or to declare approximate advances, is dramatically reduced. This corresponds to SDG Target 16.5, "Substantially reduce corruption and bribery in all their forms,"[73] and SDG Target 16.6, "Develop effective, accountable and transparent institutions at all levels."[74]

Another blockchain application to monitor the implementation of an agreement is the Blockchain Challenge developed by the Convention on International Trade in Endangered Species of Wild Fauna and Flora (CITES). This international agreement has for an objective to ensure that international commerce of specimens of wild animals and plants does not menace their existence.[75] To avoid the forging of paper-based trading authentication, the organization has established electronic permit systems (eCITES) and an Electronic Permit Information Exchange (EPIX) between state members for the exchange of eCITES permits. The EPIX platform will prevent the use of fraudulent permits and substantially streamline legal commerce in specimens of CITES-listed species.[76] Aware of the potential of blockchain technologies for CITES, the organization's Secretariat has launched a Blockchain Challenge to encourage research on the use of this technology to protect wild animals and plants from illegal trade. A new blockchain platform could indeed allow all transactions to become transparently available and immutable.[77] This corresponds to SDG Target 15.7, "(. . .) end poaching and trafficking of protected species of flora and fauna and address both demand and supply of illegal wildlife products (. . .),"[78] and SDG Target 15.c, "(. . .) combat poaching and trafficking of protected species (. . .)."[79]

The following section examines how blockchain applications can greatly enhance the influence of civil society on individual behaviors, hence contributing to the protection of the planet.

5. Nudging citizens

For decades, civil society organizations intended to help individuals to adopt more sustainable behaviors. The results do not match the level of their efforts. This is mainly due to the difficulty of aligning long-term interests (less pollution, sustainable use of natural resources, conservation of biodiversity, etc.) with short-term interests (comfort, speed of transportation, etc.). Blockchain technology can greatly help civil society to improve its impact on individual behaviors.

The "tragedy of the commons," a concept introduced about 50 years ago by the ecologist Garrett Hardin, is often explained through the following example: small farmers sharing a common pasture must decide how many of their sheep will graze (taking into consideration their own profit but also considering the actions of other farmers). Each farmer will attempt to maximize his or her benefits by grazing as many sheep as possible. Consequently, the large number of sheep overgraze and damage the pasture. However, each farmer would benefit more in the long run if he or she had cooperated and grazed fewer sheep. Joshua Weitz and his colleagues from the Georgia Institute of Technology conducted research to identify what scenario could avert the tragedy of the commons, and they discovered that it was possible to push individual agents to cooperate only when the incentive (to cooperate) was strong enough, even if the environment was already depleted or when other agents would continue to degrade the common resources.[80]

The environment is fragile and a complex issue to govern, and we all own a part of this global public good. The air, oceans, biodiversity, and the earth as a whole are our "commons" and our "blockchain" in a way. Each of us has the possibility to either support or diminish "(. . .) the sustainable management and efficient use of natural resources" by choosing to use a car or a bicycle, throwing everything into one bin or recycling, or buying local and organic products or highly processed food from the other side of the world. On paper, this decision is quite simple: theoretically, we would all tend to choose the option that is most favorable to the planet (if provided with adequate information). However, we often do not, although we know what the "right" choice for the planet is – and therefore for ourselves and future generations. One of the many explanations is linked to present-time cognitive bias,[81] which means that individuals prefer a smaller and earlier prize rather than a larger and later one.[82] This bias is gradual and evolves over time. Indeed, individuals have the tendency to over-emphasize the value of the smaller-sooner prize over the larger-later one. This means that they will choose the short-term reward even if the later one offers more substantial benefits, or if the short-term focus triggers longer-term losses.[83]

In concrete terms, individuals will choose the reward of driving their car now over the losses triggered by climate change in the future. This is quite irrational behavior, as one's actions go against one's self-interests. Indeed, the threat or danger in the near or distant future remains the same and should be considered equally. Other cognitive biases explain this type of irrational behavior. But this is merely to illustrate the point that we, as individuals, can sometimes behave irrationally. This means that increased transparency and trust alone will not suffice to protect oceans, fight climate change, and alleviate the extinction of biodiversity. It is also a question of incentives.

Satoshi Nakamoto understood human nature and the tragedy of the commons. He therefore designed the blockchain technology based on two incentives: reward and ownership. First, it rewards the blockchain miners who help maintain it. Second, the reward (in the case of a bitcoin blockchain) is also a share of the platform: the blockchain belongs to those who own and use the value recorded on the network.[84] These two principles imply that agents have an interest in cooperating and protecting the platform (which can be likened to "the commons") rather than forging multiple identities, diluting rights, and depreciating the trust value of the network to maximize their own gains. By acting in their own self-interest, individual agents also act in the interest of the network. The consensus mechanisms, discussed previously, ensure that users of the network have a higher incentive to cooperate than to deteriorate "the commons."

Satoshi Nakamoto succeeded not only in creating a strong incentive to cooperate (and avoid destruction of the network), but also in ensuring that each user or participant in a blockchain is unique, is authentic (although anonymity can be maintained), and has a transparent reputation based on past transactions and *Proof of Work*. Because of the amount of work and resources required to mine and generate new blocks, it is impossible for miners to forge their own identity or have multiple ones. Their interest is to pool all their resources and always increase their reputation.

Blockchain offers a unique opportunity for civil society organizations to nudge individuals toward aligning their long-term and short-term interests. Innovative blockchain solutions allow civil society to influence individual behaviors, and in that sense, contribute to the protection of the environment. First, cryptographic tokens can be issued every time a good action is taken. For instance, individuals can receive a short-term gain (cryptographic tokens) every time they recycle plastic containers, cans, or batteries for instance. Plastic Bank empowers recycling ecosystems around the world by supporting the exchange of plastic for money, items (e.g. cooking fuel), services (e.g. phone charging), and cryptographic tokens. Its objective is to reduce ocean plastics by gathering a billion people to monetize plastic waste and improve lives.[85] Another example is RecycleToCoin, which aims to provide cutting-edge incentives to recycle single-use plastic bottles and aluminum cans. Based on a blockchain mobile app, this system allows the exchange of recyclable waste for cryptographic tokens. Participating individuals can then use their tokens to finance renewable energy projects, which will produce clean electricity via solar, biomass, wind powered, and ground source systems.[86] Similarly, a bonus on a blockchain-based reputation system can enhance the image of an individual.

SolarCoin[87] is a digital currency that functions similarly to frequent flyer air miles: individuals and households producing solar electricity receive SolarCoins as a reward for each Mwh of solar electricity generated. Since solar installations are expensive to build, SolarCoin aims to reduce the payback time by offering an additional incentive to produce solar powered electricity. The SolarCoin Foundation, in partnership with the solar industry, creates a virtuous circle, which will in turn lead to an uptake of solar equipment and support new jobs in this field. These examples correspond to SDG Target 7.2, "(. . .) increase substantially the share of renewable energy in the global energy mix (. . .),"[88] and SDG Target 12.5, "(. . .) reduce waste generation through prevention, reduction, recycling and reuse (. . .)."[89]

The following section examines how blockchain applications can greatly enhance the influence of civil society on individual behaviors, hence contributing to the protection of the planet.

6. Concluding remarks

Technological innovation is an ongoing process. It never stops. Technology discoveries feed each other, triggering the emergence of other technology breakthroughs, as well as new social and organizational practices. These new behaviors, expectations, and patterns of interaction[90] also lead to new social needs and business opportunities, generating technological inventions. A stable and prosperous society is also a society where natural resources are managed inclusively and with a long-term perspective. Blockchain is not a disruptive technology offering lower-cost solutions (that would replace traditional business models and forms of organizations). It is a foundational technology in the sense that it will create new economic and social foundations. For this revolution to happen, some barriers – technological, governance, organizational, and even societal – will need to fall.[91] Pilot programs must be run to detect potential technical and security flaws. Although probably triggering an enormous impact, this technology will gradually be adopted over the coming decades, and its impact will be more diffuse.

As discussed in this chapter, blockchain offers promising solutions to finance more efficiently the protection of the environment, better monitor the environment, food security, and the implementation of projects, and nudge more effectively individuals toward more sustainable behaviors. The role of civil society in developing these innovative applications is crucial, and will place this group of actors at the center of global environmental governance in the digital age. However, they are not the only ones benefiting from the blockchain technology: individuals, businesses, and states can also embrace this technology to create a more sustainable future.

Traditionally, power electricity grids are centrally developed and managed. At the regional level, states buy and sell electricity from each other. At the national level, electricity is managed by either one or a few entities (private or public). Many natural monopolies of traditional energy producers and distributors are in place. In most cases, electricity and energy in general are questions managed mainly at the national level. If this system allows an overview of all supply and demand, it also leads to some distribution inefficiencies (in some cases surplus). Furthermore, in the parts of the world most affected by disasters, power outages are frequent and need time to be fixed (to reconnect to the rest of the power grid or to rebuild a power plant). Also, with centralized energy management strategies, the infrastructure is more vulnerable to cyber-attacks, making the state, its population, and its economy much more vulnerable, and necessitating higher levels of cybersecurity.

In the energy sector, blockchain technology can support individual and local solar power producers in selling their electricity surplus from one neighbor to another. Blockchain can indeed help reduce (and in some cases eliminate) the need to transport energy (electricity, oil, gas, etc.) over very long distances, and to store it (thanks to trading electricity among neighbors and communities). This corresponds to SDG Target 7.1, "(. . .) ensure universal access to affordable, reliable and modern energy services (. . .)."[92] Furthermore, new power plants (nuclear, coal, or gas based) are very expensive to build and to operate. They also represent a high risk for pollution security. Decentralized systems tend to prove more resilient than centralized grids in the case of natural disasters, allowing people to rely on a local network when the major infrastructure fails.

On the other hand, locally produced electricity could form a network of Independent Power Producers (IPPs), which would reduce the importance of third-party intermediaries (private sector or the state) and increase energy use efficiency. IPPs can decide to sell their excess energy for maximum profit or choose to donate a portion to a low-income area: "You don't have the billing components around it, you don't have the infrastructure losses or the accounting losses in the system," says Lawrence Orsini, co-founder of Transactive, a joint venture between ConsenSys and LO3 Energy.[93] SunContract is a platform, based on blockchain, that enables individuals and households to buy green electricity, but also solar power plants, heat pumps, and storage units.[94] Thanks to a wide array of energy products and services, people can become completely energy self-sufficient, and even choose to sell home-generated electricity. SunContract offers the possibility to produce and locally trade green electricity, bypassing traditional centralized producers and distributors of the energy sector.

The global carbon credits scheme requires approximately US$979 million annually to administrate it.[95] Blockchain and smart credits can first reduce these costs by storing all legal documents on the ledger, and setting up automatic transactions at specific dates when certain conditions are met through smart contracts. In addition, blockchain and cryptocurrencies allow funds to be transferred from one agent to another without an intermediary (and therefore at no cost). To reduce gas emissions, states could create a "carbon currency" to demystify and consolidate the carbon market. Since carbon credits are data-driven, rely on numerous approval stages, and exist independently from the physical impacts to which they are associated, they could easily become a digital currency, which would allow carbon emissions to be tracked transparently and consistently. As a consequence, states could measure, monitor, and trade carbon emissions transparently. Retailers could take into account the carbon impact of each product they sell. And consumers could evaluate clearly and easily the environmental impact of what they purchase (be it a service or a product).[96] This corresponds to SDG Target 12.6, "(. . .) Encourage companies, especially large and transnational companies, to adopt sustainable practices (. . .)."[97]

However, blockchain technology, like many other innovations, also raises a series of concerns. First, blockchain technology, and in particular bitcoin blockchain platforms, requires high levels of computing capacity and therefore a high level of electricity consumption. Although this is difficult to measure precisely, most studies confirm that this new technology poses a real question concerning the sustainability of its potential development and application to new fields of society and economy. Second, cryptocurrencies offer an opportunity for governments to develop innovative surveillance systems through the human interfaces. Bitcoin's Mt. Gox Bitcoin exchange showed fraudulent transactions, and some bitcoin wallets were hacked. Third, the question of immutable information can be problematic in some cases. Indeed, the right to be forgotten according to an EU ruling requires the removal of some webpages in specific situations. With blockchain, this would simply not be possible. Each transaction remains visible to all forever. Finally, blockchain is possible only where technology is more equally accessible. In today's world with a substantial digital divide, states will have to "(. . .) increase access to information and communications technology and strive to provide universal and affordable access to the Internet (. . .)."[98]

These concerns aside, blockchain has the potential to positively affect the protection of the planet and enhance the role of civil society in global environmental governance. This technology, by allowing civil society to develop more transparent, efficient, and safe financing mechanisms, to improve financing mechanisms, and to nudge more effectively individuals toward adopting more sustainable and environmentally friendly behaviors, positions these organizations at the center of technological change, which has the potential to radically transform global environmental governance's actors and mechanisms. The routine type of administrative functions that used to be necessary to coordinate and finance field projects will be gradually replaced by blockchain-based applications. It will allow for more funding and more human capacity to be dedicated to the protection of the environment. This change will probably favor civil society, since these actors are already recognized as field and knowledge authorities.

Notes

1 Schwab, K., 2015. The fourth industrial revolution: What it means, how to respond. *Foreign Affairs*. www.foreignaffairs.com/articles/2015-12-12/fourth-industrial-revolution
2 Although many different variations of distributed multi-ledgers exist, the term "blockchain" tends to be generally accepted as an umbrella term to describe all of them.
3 Witte, J.H., 2016. The blockchain: A gentle four page introduction. *Cornell University Archives*. http://arxiv.org/abs/1612.06244
4 Tapscott, D., Tapscott, A., 2016. The impact of the blockchain goes beyond financial services. *Harvard Business Review*. https://hbr.org/2016/05/the-impact-of-the-Blockchain-goes-beyond-financial-services
5 Nakamoto, S., 2008. *Bitcoin: A peer-to-peer electronic cash system*. https://bitcoin.org/bitcoin.pdf
6 See Chapters 2 (websites), 3 (social media platforms), and 4 (internet and multi-stakeholder participation) of this book.
7 UNFCCC, 2018. *UN supports blockchain technology for climate action*. See more at https://cop23.unfccc.int/news/un-supports-blockchain-technology-for-climate-action
8 Since 2010, this team of experts from multiple scientific and policy backgrounds reviews every year 100 potential issues and highlights 15 of them.
9 Sutherland, W. et al., 2017. A 2017 Horizon scan of emerging issues for global conservation and biological diversity. *Trends in Ecology and Evolution*, 32(1), p. 31.
10 Hereby called indifferently "platform" or "network."
11 Hereby called indifferently "members," "participants," or "agents."
12 Baliga, A., 2017. *Understanding blockchain consensus models*. www.persistent.com/wp-content/uploads/2017/04/WP-Understanding-Blockchain-Consensus-Models.pdf
13 Tapscott, D., Tapscott, A., 2017. *Blockchain revolution: How the technology behind Bitcoin is changing money, business, and the world*. New York, NY: Penguin Books and Random House, p. 35.
14 Iansiti, M., Lakhani, K.R., 2017. The truth about blockchain. *Harvard Business Review*. https://hbr.org/2017/01/the-truth-about-Blockchain
15 Witte, J.H., 2016. The blockchain: A gentle four page introduction. *Cornell University Archives*, p. 2. http://arxiv.org/abs/1612.06244
16 Di Pierro, M., 2017. *Op Cit.*, p. 93.
17 Witte, J.H., 2016. *Op Cit.*, p. 1.
18 Castilla-Rubio, J.C., Robins, N., Zadek, S., 2016. *Op Cit.*, p. 13.
19 Castilla-Rubio, J.C., Robins, N., Zadek, S., 2016. *Fintech and sustainable development – assessing the implications*. Nairobi: UNEP Inquiry.
20 UN Environment has also published numerous reports on this question over the last few years. See in particular the 2015 and 2016 editions of "The Financial System We Need:

Aligning the Financial System with Sustainable Development" by UNEP Inquiry. http://unepinquiry.org/publication/inquiry-global-report-the-financial-system-we-need/

21 Botsman, R., 2015. The changing rules of trust in the digital age. *Harvard Business Review*. https://hbr.org/2015/10/the-changing-rules-of-trust-in-the-digital-age

22 Dow Jones Sustainability Index. For more information. www.sustainability-indices.com

23 United Nations Principles for Responsible Investment. For more information. www.unpri.org

24 Global Reporting Initiative. For more information. www.globalreporting.org/standards

25 Di Pierro, M., 2017. what is the blockchain? *Computing in Science & Engineering*, 17, p. 93.

26 Baynham-Herd, Z., 2017. Technology: Enlist blockchain to boost conservation. *Nature*, 548, p. 523. www.nature.com/articles/548523c

27 Till, B.M., Afshar, S., Peters, Alex W., Meara, J.G., 2017. Blockchain and global health. *Foreign Affairs*. www.foreignaffairs.com/articles/world/2017-11-03/Blockchain-and-global-health

28 SDG Target 1.4, "By 2030, ensure that all men and women, in particular the poor and the vulnerable, have equal rights to economic resources, as well as access to basic services, ownership and control over land and other forms of property, inheritance, natural resources, appropriate new technology and financial services, including microfinance." See more here https://unstats.un.org/sdgs/indicators/indicators-list/

29 SDG Target 15.6, "Promote fair and equitable sharing of the benefits arising from the utilization of genetic resources and promote appropriate access to such resources, as internationally agreed." See more here https://unstats.un.org/sdgs/indicators/indicators-list/

30 SDG Target 10.c, "By 2030, reduce to less than 3 per cent the transaction costs of migrant remittances and eliminate remittance corridors with costs higher than 5 per cent." See more here https://unstats.un.org/sdgs/indicators/indicators-list/

31 Zachary Baynham-Herd, 2017. Technology: Enlist blockchain to boost conservation. *Nature*, 548, p. 523. www.nature.com/articles/548523c

32 SDG Target 13.b, "Promote mechanisms for raising capacity for effective climate change-related planning and management in least developed countries and small island developing States, including focusing on women, youth and local and marginalized communities." See more here https://unstats.un.org/sdgs/indicators/indicators-list/

33 SDG Target 15.1, "By 2020, ensure the conservation, restoration and sustainable use of terrestrial and inland freshwater ecosystems and their services, in particular forests, wetlands, mountains and drylands, in line with obligations under international agreements." See more here https://unstats.un.org/sdgs/indicators/indicators-list/

34 SDG Target 15.c, "Enhance global support for efforts to combat poaching and trafficking of protected species, including by increasing the capacity of local communities to pursue sustainable livelihood opportunities." See more here https://unstats.un.org/sdgs/indicators/indicators-list/

35 BitGive. For more information. www.bitgivefoundation.org

36 BitHope. For more information. https://bithope.org

37 See the Natural Capital Finance Alliance. www.naturalcapitaldeclaration.org/about

38 Ibid.

39 Castilla-Rubio, J.C., Robins, N., Zadek, S., 2016. *Op Cit.*, p. 13.

40 See the Forest Carbon Partnership Facility. www.forestcarbonpartnership.org/

41 REDD+ is the acronym for countries' efforts to Reduce Emissions from Deforestation and forest Degradation, and foster conservation, sustainable management of forests, and enhancement of forest carbon stocks.

42 Castilla-Rubio, J.C., Robins, N., Zadek, S., 2016. *Op Cit.*, p. 13.

43 For more information. www.iucn.org/theme/protected-areas/our-work/iucn-green-list

44 For more information. https://gls.porini.foundation/en/

45 This is an initiative from IUCN and the Swiss-based Porini Foundation.
46 EcoChain. For more information. www.bcdc.online/ecochain
47 Tapscott, D., Tapscott, A., 2017. *Op Cit.*, p. 50.
48 SDG Target 7.2, "By 2030, increase substantially the share of renewable energy in the global energy mix." See more here https://unstats.un.org/sdgs/indicators/indicators-list/
49 Baroke, S., 2017. Passport ethical labels – key findings. *Euromonitor*. https://blog.euromonitor.com/2016/06/passport-ethical-labels-key-findings.html
50 SDG Target 12.8, "By 2030, ensure that people everywhere have the relevant information and awareness for sustainable development and lifestyles in harmony with nature." See more here https://unstats.un.org/sdgs/indicators/indicators-list/
51 Provenance. For more information. www.provenance.org
52 FoodTrax. For more information. www.foodtrax.nl
53 Everledger. For more information. www.everledger.io
54 Chapron, G., 2017. The environment needs cryptogovernance. *Nature*. www.nature.com/news/the-environment-needs-cryptogovernance-1.22023
55 SDG Target 12.4, "By 2020, achieve the environmentally sound management of chemicals and all wastes throughout their life cycle, in accordance with agreed international frameworks, and significantly reduce their release to air, water and soil in order to minimize their adverse impacts on human health and the environment." See more here https://unstats.un.org/sdgs/indicators/indicators-list/
56 SDG Target 12.3, "By 2030, halve per capita global food waste at the retail and consumer levels and reduce food losses along production and supply chains, including post-harvest losses." See more here https://unstats.un.org/sdgs/indicators/indicators-list/
57 Castilla-Rubio, J.C., Robins, N., Zadek, S., 2016. *Op Cit.*, p. 33.
58 For more information, see https://stratisplatform.com/tag/earth-twine/
59 Girard, P., Du Payrat, T., 2017. An inventory of new technologies in fisheries. *Issue Paper*, OECD, p. 14.
60 For more information, see WWF Australia website. www.wwf.org.au/news/news/2018/how-blockchain-and-a-smartphone-can-stamp-out-illegal-fishing-and-slavery-in-the-tuna-industry#gs.7WLneaI
61 Read more at https://phys.org/news/2018-01-blockchain-tuna-traceability-combat-illegal.html#jCp
62 SDG Target 14.4, "By 2020, effectively regulate harvesting and end overfishing, illegal, unreported and unregulated fishing and destructive fishing practices and implement science-based management plans, in order to restore fish stocks in the shortest time feasible, at least to levels that can produce maximum sustainable yield as determined by their biological characteristics." See more here https://unstats.un.org/sdgs/indicators/indicators-list/
63 SDG Target 14.b, "Provide access for small-scale artisanal fishers to marine resources and markets." See more here https://unstats.un.org/sdgs/indicators/indicators-list/
64 Forest Stewardship Council. Read more at https://ic.fsc.org/en/what-is-fsc
65 For more information. www.fsc-uk.org/en-uk/newsroom/id/407
66 Ibid.
67 SDG Target 15.2, "By 2020, promote the implementation of sustainable management of all types of forests, halt deforestation, restore degraded forests and substantially increase afforestation and reforestation globally." See more here https://unstats.un.org/sdgs/indicators/indicators-list/
68 SDG Target 15.3, "By 2030, combat desertification, restore degraded land and soil, including land affected by desertification, drought and floods, and strive to achieve a land degradation neutral world." See more here https://unstats.un.org/sdgs/indicators/indicators-list/
69 Di Pierro, M., 2017. What is the blockchain? *Computing in Science & Engineering*, 17, p. 93.
70 Sutherland, W.J., et al., 2017. A 2017 horizon scan of emerging issues for global conservation and biological diversity. *Trends in Ecology and Evolution*, 32(1), p. 38.

71 SDG Target 1.4, "By 2030, ensure that all men and women, in particular the poor and the vulnerable, have equal rights to economic resources, as well as access to basic services, ownership and control over land and other forms of property, inheritance, natural resources, appropriate new technology and financial services, including microfinance." See more here https://unstats.un.org/sdgs/indicators/indicators-list/

72 SDG Target 15.6, "Promote fair and equitable sharing of the benefits arising from the utilization of genetic resources and promote appropriate access to such resources, as internationally agreed." See more here https://unstats.un.org/sdgs/indicators/indicators-list/

73 SDG Target 16.5, "Substantially reduce corruption and bribery in all their forms." See more here https://unstats.un.org/sdgs/indicators/indicators-list/

74 SDG Target 16.6, "Develop effective, accountable and transparent institutions at all levels." See more here https://unstats.un.org/sdgs/indicators/indicators-list/

75 For more information. www.cites.org/eng/disc/what.php

76 For more information, see the CITES document prepared by the Secretariat in relation to agenda item 40 on electronic systems and information technologies and entitled "The CITES Blockchain Challenge: Can Blockchain prevent the use of fraudulent CITES certificates and permits?" https://cites.org/sites/default/files/eng/com/sc/69/inf/E-SC69-Inf-33.pdf

77 Cassar, G., 2017. From predictive analytics to mining virtual markets: The tech that offers hope for wildlife. *World Economic Forum.* www.weforum.org/agenda/2018/01/from-predictive-analytics-to-mining-virtual-marketplaces-the-tech-that-offers-hope-for-wildlife/

78 SDG Target 15.7, "Take urgent action to end poaching and trafficking of protected species of flora and fauna and address both demand and supply of illegal wildlife products." See more here https://unstats.un.org/sdgs/indicators/indicators-list/

79 SDG Target 15.c "Enhance global support for efforts to combat poaching and trafficking of protected species, including by increasing the capacity of local communities to pursue sustainable livelihood opportunities." See more here https://unstats.un.org/sdgs/indicators/indicators-list/

80 *Game theory shows how tragedies of the commons might be averted.* www.rh.gatech.edu/news/583719/game-theory-shows-how-tragedies-commons-might-be-averted

81 Crawford, V.P., 2014. Now or later? present-bias and time-inconsistency in intertemporal choice. procrastination talk. *University of California Santa Barbara.* http://econweb.ucsd.edu/~vcrawfor/ProcrasOxIntertemporalSlides14.pdf

82 Fang, H., Silverman, D., 2005. Distinguishing between cognitive biases: Beliefs vs. Time discounting in welfare program participation. *University of Pennsylvania.* http://economics.sas.upenn.edu/~hfang/publication/cognitivebias/FangSilvermanBP3.pdf

83 Ibid.

84 Tapscott, D., Tapscott, A., 2017. *Op Cit.*, p. 35.

85 Social Plastic. For more information. http://socialplastic.org/ Plastic Bank. For more information. www.plasticbank.org/what-we-do/

86 RecycleToCoin. For more information. www.bcdc.online/recycletocoin

87 SolarCoin. For more information. https://solarcoin.org/

88 SDG Target 7.2, "By 2030, increase substantially the share of renewable energy in the global energy mix." See more here https://unstats.un.org/sdgs/indicators/indicators-list/

89 SDG Target 12.5, "By 2030, substantially reduce waste generation through prevention, reduction, recycling and reuse." See more here https://unstats.un.org/sdgs/indicators/indicators-list/

90 What Don Tapscott identified as collaboration, participation, and transparency in one of his previous books: Tapscott, D., 2008. *Wikinomics: How mass collaboration changes everything.* London, UK: Atlantic Books.

91 Iansiti, M., Lakhani, K.R., 2017. The truth about blockchain. *Harvard Business Review.* https://hbr.org/2017/01/the-truth-about-Blockchain

92 SDG Target 7.1, "By 2030, ensure universal access to affordable, reliable and modern energy services." See more here https://unstats.un.org/sdgs/indicators/indicators-list/

93 Rutkin, A., 2016. Blockchain-based microgrid gives power to consumers in New York. *New Scientist.* www.newscientist.com/article/2079334-Blockchain-based-microgrid-gives-power-to-consumers-in-new-york/

94 SunContract. For more information. https://suncontract.org

95 Future Thinkers, 2017. *7 ways the blockchain can save the environment and stop climate change.* http://futurethinkers.org/Blockchain-environment-climate-change/

96 Walker, L., 2017. *This new carbon currency could make us more climate friendly.* www.weforum.org/agenda/2017/09/carbon-currency-Blockchain-poseidon-ecosphere/

97 SDG Target 12.6, "Encourage companies, especially large and transnational companies, to adopt sustainable practices and to integrate sustainability information into their reporting cycle." See more here https://unstats.un.org/sdgs/indicators/indicators-list/

98 SDG Target 9.c, "Significantly increase access to information and communications technology and strive to provide universal and affordable access to the Internet in least developed countries by 2020." See more here https://unstats.un.org/sdgs/indicators/indicators-list/

6 Big data and environmental civil society organizations

1. Introduction

We live in a world of data. Today more than ever, data are at the center of the economy. Individuals, organizations, and governments consume and produce large amounts of data. The generalization of information and communication technologies (ICTs) throughout the world, combined with the increased use of mobile devices, has led to the production of vast amounts of data. Amazon, Apple, Facebook, Google, Twitter, and Netflix are among the most well-known examples of companies benefiting from petabytes of mostly unstructured data.[1] In 2011, global digital information accounted for one trillion GB, and it is expected to reach 50 trillion in 2020.[2] As a point of reference, 1 petabyte corresponds to more than 16,000 64-GB iPhone Xs.[3]

Big data are increasingly recognized as a new paradigm equal in measure to the Industrial Revolution.[4] New business models, services, and products have been developed thanks to the information extracted from data; this has become an unprecedented source of innovation and wealth. Although there is no common definition of the concept, big data refer to the new technologies that allow the collection, storage, and analysis of data sets too large for traditional processing systems.[5] The applications are endless. Big data are now common place, be it for marketing research or healthcare. Organizations can consider data sets on a global scale, which enables them to know their target audiences as never before, and advertise precisely based on their segments' needs. Political leaders can design new communication campaigns based on what they know about their potential voters: what they think, like, and dislike. For instance, the 2012 Obama campaign is considered the first big data political campaign. It collected data on a national scale using smartphones and sent out instructions to its staff and volunteers in the field based on what had been analyzed.[6] It enabled the Obama team to simulate the election 66,000 times every night based on the data collected and adapt resources accordingly. It also allowed the team to better target voters and more effectively buy TV advertising.[7]

If the concept of big data aims to describe how large and complex data sets are analyzed.[8] it also describes how organizations collect data from new sources. This process was once restricted to governments, but individuals now have their personal data and online behaviors scrutinized and analyzed by a large range of stakeholders around the globe. The massive and global adoption of social media platforms has led to an equally massive production of data, i.e. large sets of data ranging from browsing history, to the number of likes on Facebook, or the content of tweets. However, big data do not only stem from social media. The digital trail combines all the data that individual users leave behind when online. It is also combined with other sources of data, such as online and offline credit-card purchases, geo-localization of pictures taken with a smartphone, or connection to a Wi-Fi service, among many others. Since individuals are constantly connected to the internet, all daily actions are collected, stored, and analyzed by some stakeholders somewhere at some point. Most of these data are simply collected and not transformed or analyzed. This is dark data. But for the data that are collected, many ethical and legal questions have been raised. The recent General Data Protection Directive (GDPR) of the European Union intends to address these issues in order to better protect the privacy and confidentiality rights of European citizens.

"Big data goes green"[9] was the title of a recent article of the magazine *Nature*. Big data hold many promises for studying and protecting the environment, and provide unprecedented insights to support decision-making processes in a data-informed policy era.[10] Among the promises, big data are said to transform cities[11] into smart cities, as well as help societies to better overcome climate change vulnerability.[12] In terms of science, big data analysis is expected to solve a variety of environmental issues[13] from different spatial and temporal perspectives. For instance, by combining climate and health data, new models can be built to analyze the influence of climate change on people and communities.[14] Another case is using big data to examine multiple environmental stressors in marine ecosystems.[15] In addition, by collecting and processing social and biophysical data,[16] other research can focus on how humans interact with their environment, which supports decision and policy makers in different fields such as conservation,[17] air pollution abatement,[18] and urban planning.[19] Thanks to the automated data collection capacity, combined with the analysis of data-driven methodologies, voluminous data sets are captured in dynamic and remote environments.[20] This leads to a far more precise and up-to-date understanding of the earth-atmosphere system and how it relates to human activities on a global scale.[21]

Evidence-based policy making, empowered by big data, is where organizations utilize large data sets to understand complex situations (climate for instance) and design better prediction models for hurricanes and natural resource management. Furthermore, with numerous data points from the field, organizations can have a real-time view of what is happening in real life. In other words, big data do not merely allow better predictions for the future based on large data sets from the past; they also allow the present picture to be envisaged. This enables organizations to adapt their actions (short-term decisions) in real time and make better medium- and long-term policies. In terms of lobbying and advocacy, big data can also prove to be useful. Indeed, they can provide evidence and facts to support civil society when lobbying governments and decision makers to develop sustainable production and distribution processes.

This chapter provides an overview of the current and potential big data applications developed by civil society in the context of global environmental governance. Although the applications are various and numerous, an exhaustive list of all applications of these emerging technologies is difficult to produce. Therefore the chapter will focus on the use of big data to produce new knowledge. Indeed, this corresponds to one of the key roles of civil society, which is widely recognized as a scientific authority and knowledge broker. This technology enables as well a stronger cooperation between actors of civil society (citizens, scientific organizations, and CSOs for instance) and encourages new partnerships between civil society and the private sector. On the one hand, scientists, academia, and CSOs can benefit largely from the data collected by citizens in the field. On the other hand, civil society needs the technological expertise of the private sector to benefit the most from this technology. This chapter will also relate some of big data's applications to the Sustainable Development Goals (SDGs), showing how this technology can help achieve some of these goals and associated targets.

2. Big data explained

Big data are not new. In the context of the environment, data have already been "big" for decades, although there is no exact date to identify when data can be considered "big data." Since the 1980s, large data sets have been used to better understand global climate change, and have led to political and societal awareness of this problem.[22] Data, the raw information collected from various sources, have been used by scientists and organizations for centuries. But modern big data imply a level of complexity unforeseen not only in terms of the data collected, but also in terms of the intended results[23] stemming from the analysis of these large sets of data. The concept of big data was first coined in 1997 to describe the "(. . .) Challenge for computer systems: data sets are generally quite large, taxing the capacities of main memory, local disk, and even remote disk. We call this the problem of Big Data."[24] A couple of years later, three characteristics were identified to describe big data: velocity, variety, and volume (3 Vs)[25] and then value, validity, veracity, and visibility.[26]

First, big data technologies allow the collection and storage of large volumes of data. Thanks to the increasing digitalization of business and administrative processes, a wide range of new information sources has become accessible and collectable. Furthermore, a large number of individual user devices such as smartphones and tablets allow more data to be collected on a large scale. Also, open-source initiatives of public administrations allow organizations and individual users to benefit from new sources of information and produce new services and solutions heretofore unforeseen. Today, the data collected are reaching new dimensions. They are measured in terabytes, petabytes, and exabytes (5 exabytes of data correspond to every word ever spoken on the planet).[27]

Second, big data technologies allow the collection and storage of a broad variety of data. The nature of data collected is considered here. Structured data stem from sales and accounting for instance, and unstructured data stem from email or the Internet of Things (IoT) for instance. Big data refer to uninterrupted and frequently simultaneous streams of data stemming from several sources in terms of origin, type, time, and space. For instance, satellite data are combined with photos taken by social media users to help conservation work.[28] The combination of social and physical data is a key feature of big data.

Data originate from three types of sources. Transaction data are often the traditional source of structured data. These data stem from internal information systems such as enterprise resource planning systems. Human data are more recent and stem from the use of social media instruments such as Twitter and Facebook, among others. Lastly, sensor data stem from sensors or the Internet of Things, which produce large sets of data on a continual basis.[29] IoT originated in the Auto-ID Centre at the Massachusetts Institute of Technology (MIT) in the late 1990s to identify the RFID[30] infrastructure.[31] Today it describes a network of hardware "things" that are connected to each other through the internet, allowing them to create a network of information.[32] These connected devices also communicate with users; they include washing machines, lights, CCTV cameras, RFID tags, vehicles, thermometers, nuclear reactors, air conditioning units, and many more.[33]

Big data are often real-life facts about populations, collecting all data rather than samples, and they are often messy, meaning unstructured, with limited or no quality control.[34] Furthermore, big data are in many cases collected by specialized organizations through automated technologies and without necessarily having the explicit consent of the providers of the data. This means that the data collected are often multi-purpose, non-disciplinary, multi-institutional, and multinational.[35] In terms of sources, three types can be identified: volunteered (provided by users), automated (provided automatically by a device such as a smartphone), and directed (measured by a human operator).[36] Although most big data applications for nature conservation and environmental governance concentrate on directed sources, automated, volunteered, and crowdsourced data sets represent a high potential (although with some limitations due to their nature).[37]

Third, big data technologies collect and analyze data at a high speed. The velocity of creation, capture, and processing of large sets of data is extremely relevant in times of emergency situations and natural disasters for instance.[38] Fourth, the veracity of the large sets of data is crucial. Big data technologies allow data to be collected in a systematic and automated way that ensures accuracy and integrity.[39] However, unstructured and unchecked data streams do not comply with standard quality control processes.[40] The velocity of creation, capture, and processing can also trigger some challenges in terms of data veracity and value. For instance, large data sets from air quality sensors may also come from crowdsourced data, which leads to higher levels of uncertainty[41] in terms of data viability.

Lastly, the data collected and stored are processed to make sense of them, make them visible, and produce an added value. Thanks to big data technologies, unstructured data from a wide variety of sources can be analyzed in order to make sense of it.[42] Big data analytics is recognized as one of the five main foundational technologies in terms of analytics research and describes the use of data mining and statistical analysis.[43]

Nowadays, organizations have access to vast amounts of data. The question is therefore not so much anymore about the capacity to collect data, since an infinite number of data sets are produced constantly, but rather how fast to analyze them, in order to avoid accumulating raw data (also called dark data). Although there is general agreement that many organizations can benefit from collecting large amounts of data, it sometimes remains unclear what the concrete applications can be and the potential outcomes of the processing and analysis of these data sets.[44] Indeed, the complexity of data collected can lead to an unlimited number of applications, although often the outcomes are unknown at the present time. This explains why many organizations collect all data possible, not only the relevant data for a current application, since in years to come, the irrelevant data might become a source of innovation and wealth. It is a real challenge for most organizations to understand not only what they wish to know today, but what they could develop for the future with unstructured and structured data. Before processing and analyzing data, the organization must indeed have an idea of what the outcome could be.

The next sections will focus on the use of big data technology to produce new knowledge. The data collected can stem from citizen environmentalists, who are able to gather large data sets thanks to mobile applications developed by civil society organizations. The collection and analysis of large sets of data can also be enabled by technologies that are developed thanks to partnership with the private sector, and in collaboration with the scientific community.

3. Citizen crowdsourcing

Crowdsourcing refers to the involvement of citizens in a governmental or administrative decision-making process. In a situation of emergency, people in the field upload photos of the disaster area, which allows better decisions to be taken in terms of resource management for instance. Similarly, residents of a city can indicate through their smartphones the location of potholes on the street; the city of Boston already started using this crowdsourcing roadside maintenance technology in 2012. Since then, several versions of its "Street Bump" app have been launched, each one becoming more and more precise in detecting potholes and helping the city administration to better allocate its resources to fix the roads, all with the help of citizens.[45] Many other instruments also provide decision makers with solid situational awareness: social media, sensors, environmental data, geographic information systems, analytical algorithms, and prediction modeling, among others.[46]

Citizen science describes the involvement of the general public in the processes of scientific observations. This includes tracking the migration of various species, illustrating the evolution of specific ecosystems, and becoming the eyes and ears of scientists in the field, who could not, due to a lack of resources and technology, monitor all these changes and evolutions at local and global levels. Some technologies are specifically dedicated to the participation of citizens in scientific observations. The big data they create are then collected, stored, and analyzed by scientists, who can improve their research with real-time data.

FieldScope is an example of one of these technologies. It allows citizen scientists to easily upload their data and visualize them on maps and graphs.[47] They are then published online, where users can explore interactive maps and data. Citizens can also conduct fieldwork and share observations and stories, as well as participate in social and scientific networks to document species population abundance and distributions.[48]

Thanks to this technology, the Chesapeake Bay Watershed Project enables community members and students to analyze water quality at local and regional levels, and then share their results with scientists and the general public in order to take action.[49] FrogWatch USA was launched by the Association of Zoos and Aquariums to enable citizens to participate in the effort of protecting amphibians through reporting data on the calls of frogs and toads in their local environment. Their observations have created a large scale, long-term data repository on frogs and toads in the United States. Participants can also discover maps and graphs about amphibians, their abundance and distribution, timing of calls, and other data.[50]

Litterati[51] is a very good example of leveraging crowdsourced data. This startup is building a data set to identify what kind of litter is where, how much of it there is, and how it evolves through time. Since litter is everywhere, its impact on the environment is dramatic. This startup is based on a community of citizens who take photos of litter wherever they are. It is then tagged with key words that allow identification of the most common brands, geolocated and time-stamped, thereby permitting the creation of littering profiles for cities and landscapes. This helps public administrations and companies to become more efficient and effective in their cleaning and recycling efforts.

The BirdReturns[52] program was developed by The Nature Conservancy (TNC) to protect the migration of birds between Canada and Mexico. This Pacific Flyway is threatened due to the increasing loss of wetlands in California's Central Valley. Cornell University (USA) is gathering geotagged photos of birds to include in a shareable database. Amateur birders, who as a hobby take photos throughout the year of birds in the field, have revealed themselves to be extremely useful. Instead of writing down the information on a piece of paper or publishing the photo in a specialized journal, they can use an application called eBird developed by Cornell's Lab of Ornithology, which automatically adds to the database the photo of the bird, as well as the time and location of the photo. With the name of the bird, Cornell's Lab of Ornithology can determine precisely where and when birds fly between Canada and Mexico. But not only this. In collaboration with TNC, it developed a computational model that can precisely predict when and where the birds would be in any given season.[53] This then allows TNC to collaborate with rice farmers to create pop-up wetlands, where farmers can either add water to their fields in advance or leave it longer than usual, so that birds can rest and feed in these temporary wetlands.[54] Thanks to big data, citizens, an NGO, an academic institution, and rice farmers collaborated to save the Pacific Flyway.

Cornell's Lab of Ornithology has also developed Project FeederWatch, which enables the university to gather data from thousands of people in the United States who count birds at their feeders. It enables scientists to apprehend long-term trends in bird population abundance and distribution. The data indicate which bird species are located in specific parts of the North American continent every season, and how many of each species are observed, allowing scientists to produce a very precise bird population map. It allows them to assess the influence of non-native species on native bird communities, discover the association between birds and habitats, and track movements of birds in winter.[55] The big data collected by FeederWatchers allow scientists to apprehend the timing and extent of winter irruptions of winter finches and other bird species, increases or reductions in the winter ranges of feeder birds, the types of foods and environmental dynamics that attract birds, and how disease is spreading among birds showing up at feeders. Thanks to this information, scientists can understand what causes bird extinction and act upon it before it is too late.[56]

NestWatch is another citizen crowdsourced project developed by Cornell's Lab of Ornithology. It is building a large database with information about nest location, habitat, bird species and number of eggs, and number of young, thanks to the participation of citizens throughout North America. The NestWatch Mobile App records what citizens capture. Their observations are then collected and compiled with thousands of others. This growing database is used by researchers to monitor the breeding success of birds. But it also allows them to combine these data with historical data, to better comprehend how breeding is affected by other environmental factors such as climate change, urbanization, and habitat degradation and loss. Without the help of citizens, it would be impossible to gather sufficient information to precisely track nesting birds.[57]

The four-day Great Backyard Bird Count (GBBC) was launched in 1998 by the Cornell Lab of Ornithology and National Audubon Society, and is the first digital citizen-science project to collect data on wild birds and to display results in near real time. Today, about 160,000 people from 120 countries in the world participate in the four-day event each February to produce an annual snapshot of the distribution and abundance of more than half of the world's species.[58]

The African Wildlife poisoning database was developed by the Vulture Specialist Group of the IUCN Species Survival Commission.[59] The objective is to gather data on past and ongoing incidents of wildlife poisoning in order to evaluate not only the breadth and impact of this issue on the endangered species targeted by the poisoning, but also the unintentional effect it has on any species scavenging on poisoned carcasses, which in turn leads to a further negative impact on the ecosystem and local communities. Citizens can take a photo of the incident and upload it on the website through an easy-to-use interface. It then generates an interactive map of all poisoning events throughout Africa.[60]

Conserve.io is a mobile app technology that allows experts and conservationists to collect and visualize big data. On the one hand, it helps users to collect sets of observation data over a pre-defined "track" or trip. Their data are collected either online or offline, and then automatically synchronized with cloud-based databases. Spotter, its data collection platform, also allows users to manage data sets and produce reports. On the other hand, this technology allows managers to relay real time, GPS-triggered management boundaries, seasonal openings and closings, and compliance notices to the data collectors at any time; they then become the eyes and ears for an organization's conservation and resource management efforts.[61] Conserve.io is used by many initiatives to drive citizen science, collect and visualize data, and protect nature. Whale Alert is a large network of CSOs, governmental agencies, and shipping and technology companies dedicated to decreasing deadly ship strikes of whales.[62] Another initiative is the Manatee Alert, which informs sailors and boaters about speed areas and critical manatee natural habitat locations.[63] SharkNet is a collaboration between Stanford University and TOPP Shark Institute. Their application is designed to improve the visibility of conservation work by showing the near real-time positions of a broad number of great white sharks in the "Blue Serengeti" region, well-known for its abundant marine life.[64]

Bumble Bee Watch is another citizen-science project that aims to monitor and protect bumble bees in North America. Volunteers take and upload photos of bumble bees, which are then verified by experts and added to the website. Since they are extremely distributed throughout space, bumble bees are difficult to monitor. Thanks to citizen conservationists, it is possible to identify remnant populations of rare species and take action to protect them before they become extinct. Through this citizen engagement, scientists can determine the status and conservation needs of bumble bee populations.[65] On its website, it is also possible to visualize all data collected by citizens.

The next section will discuss how the need for additional data motivates civil society organizations (CSOs) to develop new partnerships with the private sector. Hence, these new partnerships enable CSOs to benefit from the latest technological innovations, and produce new scientific knowledge. In this context, big data allow for innovative forms of collaboration between academia, governments, civil society, and the private sector, helping reach SDG Goal 17 entitled "Revitalize the global partnership for sustainable development."[66]

4. Partnerships with the private sector

In 2013, Conservation International (CI) and Hewlett Packard Enterprise (HPE) collaborated to launch HPE Earth Insights to apply big-data-processing technology to track biodiversity in tropical forests. The idea is based on early detection of environmental challenges that threaten our survival. Millions of photos of species are captured via camera traps, which enables scientists to monitor how land use, human activity, and climate change impact species abundance and distribution.[67] Data are collected by the Tropical Ecology Assessment and Monitoring (TEAM) Network, comprising CI, the Smithsonian Institution, and the Wildlife Conservation Society. Thanks to HPE big data analytics technology, HPE Earth Insights generates early warnings to better allocate conservation efforts, as a result of near real-time analytics about tropical forests at 16 locations across 15 states. Among the first results, 14% of the populations of the 275 observed species are either significantly declining or likely to be declining.[68]

The Great Elephant Census (GEC) is a big data project led by Microsoft co-founder Paul Allen's company Vulcan Inc. and a wide range of civil society organizations including Elephants Without Borders. The project is also supported by other organizations and individuals on the ground in Africa and globally including African Parks, Frankfurt Zoological Society, Wildlife Conservation Society, The Nature Conservancy, IUCN African Elephant Specialist Group, Howard Frederick, Mike Norton-Griffith, Kevin Dunham, Chris Touless, and Curtice Griffin.[69] Conservation efforts are very complicated without up-to-date data about elephant populations. The objective of this project is therefore to provide accurate and reliable data about elephant population abundance and distribution in Africa. The Great Elephant Census enables researchers to fly small planes to capture observational data of elephants and elephant carcasses. So far, a total of 352,271 elephants have been sighted in 18 countries. The resulting database offers impartial and verified information to governments, scientists, civil society, and all wildlife stakeholders to help them better manage and shelter elephant populations. In its last report, the GEC showed that the population of Savanna elephants decreased by 30% between 2007 and 2014, and that the current rate of decrease is 8% per year. Of the elephants surveyed, 84% were located in protected areas, but a large number of carcasses were also spotted in these areas, showing that elephants are struggling both inside and outside parks.[70]

The Nature Conservancy developed a cloud-based application called PAM (a Portuguese acronym for Municipal Environmental Portal App) to: (1) allow land managers to monitor and reach their environmental goals, and (2) ensure that landowners comply with Brazil's forest laws. It is based on the idea that accurate monitoring leads to better enforcement, and therefore a reduction in illegal deforestation. However, it was not widely used. Thanks to its cooperation with IBM's nonprofit Corporate Services Corp, it worked to overcome some of the technical barriers that prevented its wide adoption, among which were integrating with other environmental management systems, providing access to the internet in remote areas, and developing new features including mobile platforms.[71]

In Indonesia, Greenpeace, the Union of Concerned Scientists, and the Rainforest Alliance developed the High Carbon Stock Approach (HCSA),[72] in collaboration with a large range of stakeholders including consumer packaged goods (CPG) companies and commodity producers. HCSA utilizes big data technology to differentiate between the most valuable forests that need protection, and the areas with low carbon or biodiversity value, which can be designated to harvest commodity crops. HCSA helps companies to ensure they do not support deforestation by identifying the most suitable land for crop production, and sourcing raw material, such as palm oil or rubber, that was sustainably produced. HCSA provides a wide variety of information including carbon emissions, biodiversity richness, vegetation density, land-source planning, and land rights. Companies that use HCSA include Asia Pulp and Paper (APP), CPG companies such as P&G and Unilever, and palm oil producers such as Wilmar, Musim Mas, and Agropalma. Local governments can also use HCSA data to enforce the law and prevent illegal logging and tree-felling.[73]

In South Africa, Intel collaborated with cloud service company Dimension Data, Madikwe Conservation Project, and i-Detect to develop a new portable technology to protect rhino populations. An ankle collar that consists of a credit-card sized Galileo board with 3G communication, storage features, and a solar panel is attached to the white rhino. It provides the geo-location and movement data of each animal. These data are encrypted and then sent to the cloud. In addition, an RFID chip is attached to its horn, which confirms that the white rhino is safe. When there is a break in communication between the RFID chip and the Galileo board, meaning that the horn of the rhino has been detached, anti-poaching teams are alerted and helicopters and ground-based vehicles set off in pursuit of the poachers. An upcoming phase should also include drones, and a new version of the wearable will be able to measure the rhino's vital statistics, such as temperature, heart rate, and traveling speed. This will allow anti-poaching teams to detect a stressed rhino and intervene before poachers catch the animal. A first round of implementation was conducted on a small scale of five animals, with promising results and at an affordable cost.[74] In that context, big data support the achievement of SDG Goal 15 seeking to "Protect, restore and promote sustainable use of terrestrial ecosystems, sustainably manage forests, combat desertification, and halt and reverse land degradation and halt biodiversity loss."[75]

The Lighthill Risk Network includes a large range of corporate, academic, and governmental entities that apply big data technologies to issues related to risk management and natural disasters. It collaborates in particular with the Institute for Environmental Analytics,[76] a UK data research center. Also, the University of Reading opened a new multi-million-pound world-class environmental science "big data" research and analysis center in 2015.[77] Its objective is to support decision makers by providing cutting-edge knowledge and scientific evidence about a large range of global issues, including global food security, climate change, and extreme weather, among others.[78]

The next section will discuss how academia and the scientific communities have embraced big data technologies to collect large data sets, analyze them, and produce new scientific knowledge.

5. Scientific community

Geographic Information Systems (GIS) are information systems that allow the collection, storage, analysis, and presentation of spatial and geographical data[79] through maps and imagery. Since 60% of big data contain position information,[80] GIS make extensive use of big data to analyze geographical areas and support decision makers. An important contribution that big data make to GIS is for spatial analytics. For instance, the use of combined data from large sensor networks and from social media platforms allows all phases of disaster management to become more efficient.[81] Disaster managers and decision makers have access to real-time information from the field through social media, and can cross-reference these data with satellite data and aerial imagery, to have a more complete view of a hurricane or flood situation.

The Climate and Environmental Monitoring from Space (CEMS) platform was launched in 2012 by UK academic and industrial partners with the objective of offering climate and Earth Observation (EO) data and services. Among its applications, it has generated global data sets related to climate observation, analyzed satellite data, and created new algorithms and products that associate EO with other environmental data sets.[82] Some of its applications include forest carbon stock monitoring (Rezatec), weather and climate analysis for agricultural production (WeatherSafe), and habitat monitoring and assessment (GeoSeren).[83] Hence, big data technology supports the development of accurate climate change models, and helps prepare for climate-related issues and disasters, in support of SDG Goal 13 to "Take urgent action to combat climate change and its impacts."[84]

Thanks to the constant improvement of computational capacity and with the generalization of machine-learning algorithms, researchers and policy makers can better understand climate change. JASMIN for instance, a facility run by the Centre for Environmental Data Analysis (CEDA), which combines a supercomputer and a data center, is doubling its storage capacity to reach 44 petabytes of information, equivalent to ten billion photos.[85] This change was necessary to adapt to the rapid increase in earth and environmental science data generated by a wide variety of tools located in the air and in space (e.g. the European Space Agency's (ESA) series of Sentinel satellites produces terabytes of data each day, reaching 10 petabytes per year), but also on land and in the sea.[86] JASMIN compiles data from multiple research groups and the European Space Agency, among others, and from climate simulations produced by supercomputers in Europe. It allows researchers to experiment with various models and ideas in relation to these simulations. For instance, researchers from China, South Korea, and Sweden discovered that desertification of arable lands takes place when global temperatures increase by about 1.5 to 2 degrees Celsius.[87]

To make the most out of big data, the first principle is to be able to share data sets with other research groups and organizations. Although most data centers make their data sets available to everyone, data sharing can be a challenge, due to formatting specificities at each facility. There is no universal standard for earth science data, although some best practices and recommendations have been drawn up that substantially improve data compatibility. For instance, Stanford University used big data emanating from Africa to find out what regions have higher living standards than others. The basic assumption was that these areas would have electricity, and therefore light at night. By applying machine learning to satellite images from five African countries, the university could identify which regions had the most developed electricity infrastructures and therefore which ones were the most and least developed.[88] Another example from Wuhan University in China showed that by combining satellite remote sensing and social media messages from the same area, in this case Shanghai, the university could more accurately present not only the rapid urban development of this area, but also the related human activities, which is useful for policy and decision makers.[89]

Thanks to big data, and in particular to the combination of the Eora global trade database (documenting five billion supply chains, linking 15,000 sectors in 187 countries)[90] and The IUCN Red List of Threatened Species™ database, two scientists were able to connect the consumption of goods and services to the extinction of endangered species. Indeed, the protection of biodiversity must be tackled not only in the field, in a species' habitat, but also with a broader perspective that takes into consideration the origin of the pressure, in other words what causes the extinction. Thanks to big data, scientists are now able to assess where the pressure comes from for each biodiversity threat hotspot. This will allow policy makers, but also companies and consumers, to make more informed decisions.[91]

CANAPE, which stands for categorical analysis of neo- and paleo-endemism, was developed by the University of California Berkeley to identify the priority areas to protect as natural reserves and to help scientists apprehend the evolution of life on the planet. This new model collects and analyzes large data sets including the growing mass of data from recently digitized museum collections. Since it is not possible to protect all parts of the planet, it is essential to identify the areas most in need of protection based on their biodiversity value. This model not only assesses conservation reserves but also identifies complementary areas where biodiversity is in dire need of conservation. Thanks to intense computer calculations, it evaluates the standard measure of biodiversity, meaning the number of species located in a designated area as well as the evolution among species and their geographic rarity, or endemism. Instead of counting the number of species in a specific location, this new model focuses on relative phylogenetic endemism, meaning how species are connected with each other, their lineage, and common ancestors. So far, phylogenetic information has not been fully considered when assessing biodiversity. This new model gives more weight to species that are restricted in range. It can also identify zones with clusters of new, emerging species (neo-endemics) and zones with clusters of unique, but endangered, species (paleo-endemics).[92]

SDG Goal 14 entitled "Conserve and sustainably use the oceans, seas and marine resources" can be vastly enhanced thanks to big data capacity to collect large volumes of data stemming from wildlife abundance and distribution, allowing for a better understanding of migration, threats, and adaptation to current challenges such as overfishing, climate change, and ocean acidification, to name only a few.[93]

Conservationists, biologists, engineers, computer scientists, and educators collaborated to create an international, multi-disciplinary program to apprehend the behavior of marine life and more precisely ocean animals. The Global Tagging of Pelagic Predators (GTOPP) allows users to visualize and interact with animal monitoring data, as well as oceanographic data sets and marine life observations. This initiative stems from the Tagging of Pacific Predators and Census of Marine Life, which gave millions of animals trackable sensors to better examine their behavior. GTOPP combines these large sets of data and analyzes them to have a better understanding of open ocean ecosystems. Tags are attached to marine animals that in fact become ocean sensors, generating millions of data records that can further support scientists in understanding what factors influence animal behavior in open oceans, and also help climate scientists. Thanks to these tracking devices and techniques, GTOPP can gather, process, and display tracking data, which are accessible to the global research and educational community.[94]

6. Concluding remarks

Big data are used today in a wide range of applications in global environmental governance. This chapter focused on the production of knowledge, which is one of the main roles of civil society on the international stage. Thanks to data stemming from citizen crowdsourcing, CSOs can identify trends and patterns in the complexity of human behaviors and the diversity of species and ecosystems.[95] Moreover, through their partnerships with the private sector, CSOs can benefit from complex and pricy technological innovations, which allows them to gather and analyze new data sets that were not even possible a few years ago. The scientific community is also increasingly using big data, and a large number of universities and academic centers use these new technological tools to better monitor the planet and understand how species and ecosystems evolve, and how to better protect them. The impact of these innovations is not limited to science. Indeed, one of the roles of civil society is to influence the development of rules on the national and international stages. In that context, it can be noted that big data technologies also infuse the development of future policies and laws.[96]

Big data offer promising solutions to better monitor the planet and produce new knowledge about the environment. The role of civil society in developing these innovative applications is crucial, and will place this group of actors at the center of global environmental governance in the digital age. However, these actors are not the only ones benefiting from blockchain technology: international organizations, national public administrations, and businesses also progressively embrace this technology to create a more sustainable future.

The United Nations Secretary-General launched Global Pulse, an initiative to safely and responsively harness the power of big data to overcome some of the major challenges of the 21st century, foster innovation, and support achievement of the Sustainable Development Goals (SDGs).[97] Big data technologies will allow UN agencies to better grasp the changes in human well-being and the impact of their policies on local populations in real time and on regional and global scales.[98] Big data analytics can improve their institutional capacity, develop digital public services, and support the policy cycle.[99] These technologies can also play a vital role in the context of climate change,[100] carbon emissions and air pollution,[101] the protection of endangered species,[102] cartography, surveillance, land-use planning, archaeology, environmental studies, earth observation,[103] and more.

The United Nations World Data Forum on Sustainable Development Data (UN World Data Forum) was created in 2014 further to a recommendation from the United Nations Secretary-General's Independent Expert and Advisory Group on

Data Revolution for Sustainable Development. The UN World Data Forum is a platform where information technology and geospatial information managers, data scientists, and users, as well as civil society stakeholders, meet and discuss issues and opportunities related to data and sustainability. The first UN World Data Forum was organized in 2017 in Cape Town, and it will take place annually in a different city. In addition to being a platform for dialogue among a large range of stakeholders, the UN World Data Forum aims to investigate new ways to apply data to assess global progress and support achievement of the SDGs, as well as to create innovative solutions to deliver better data for all.[104]

Environmental health organizations are increasingly using big data. Traditionally, monitoring was performed through localized environmental sampling. Thanks to computational techniques, monitoring of toxicological issues of chemicals has for instance become more precise, and the data more rapidly generated and more representative of the present situation. High throughput, data sets from multiple sources, and deeper and meta-analysis produce more relevant data that allow better decision making. The US Environmental Protection Agency (EPA) is combining throughput testing methods and computational approaches to understand the impact that the chemicals present in the environment have on human health. This allows the agency to develop new predictive models and more effective responses that were hitherto impossible to consider due to limited data.[105] This is one example of the use of big data to better monitor the environment and ensure clean water distribution and sanitation for all, in support of SDG Goal 6.[106] The EPA is conducting several big data projects.

Thanks to big data, the EPA is able to combine data from various sources and synthesize them to produce more precise predictions for areas and species with limited data. Through the Stream-Catchment project (StreamCat), the EPA is collecting large sets of data from landscape metrics for 2.6 million streams, associated catchments, and watersheds in the USA. The objective is to develop reference models that will allow comparison with future assessments and thereby more precisely identify and quantify changes in stream and watershed conditions.[107] The EPA is also identifying the human benefits of nature and more specifically of ecosystem goods and services. The EnviroAtlas project allows users to work with interactive tools and maps to visualize how they benefit from nature, ranging from access to water to security and the economy.[108] The lack of data on the toxicity of chemicals on many species is problematic since it affects individuals, populations, and communities. The EPA has developed the Web-based Interspecies Correlation Estimation (Web-ICE), an online user-friendly instrument, to allow researchers to assess the toxicity of a chemical to a species. The assessment is generated thanks to the data collected about the toxicity of the chemical on alternative species.[109]

In order to support communities in making decisions related to public health, the EPA has developed the Environmental Quality Index (EQI). The objective of the EQI is to allow researchers to identify the relationship between the quality of the environment and health criteria such as the rate of preterm births.[110] This index encompasses information from various fields such as air, water, urban, and sociodemographic space. Among many other data, it collects data on hazardous air pollutants, water impairment, waste permits, beach closures, domestic water sources, drought status, traffic safety, public transportation, road type, business environment, public housing, and socioeconomic and crime conditions.[111]

Other big data applications focus specifically on climate change modeling. Since 2010, the NASA Center for Climate Simulation (NCCS) has developed big data research projects. NCCS utilizes supercomputing, visualization, and data interaction technologies to understand climate change and produce accurate climate project models.[112] Furthermore, NASA's Earth Observing System Data and Information System (EOSDIS), part of NASA's Earth Science Data Systems (ESDS) Program, manages the 24 petabytes of NASA's earth science data stemming from various sources such as satellites, aircraft, field measurements, and other programs.[113] In 2017, it disseminated 1.3 billion files to about three million users.[114]

The private sector also benefits largely from big data technologies. With a world population that will reach nine billion in 2050, food production is increasingly putting pressure on natural resources and ecosystems. Combined with growing urbanization, climate change, desertification, and intensification of natural disasters, agriculture should capture all of our attention. Agriculture is a complex domain because weather conditions, climate, and biophysical conditions vary so greatly around the world. This, in addition to the different production systems and sizes of agricultural farms, makes it a good candidate for big data. Thanks to big data, agricultural actors have developed their capacity to capture, analyze, and visualize data.

A variety of users can benefit from big data sets. Farmers can improve their planting decisions thanks to more precise weather predictions, and can better adapt to weather changes. Agricultural firms can combine crowdsourced data with remote sensing and weather forecast information to provide more accurate insights so that farmers can increase land-use efficiency and productivity.[115] Governmental agencies can better tailor their communication to producers based on more precise information about weather and disease spreading patterns. They can also better allocate their resources to support producers in situations of hyperlocalized weather events. Donors can design development programs that are more adapted and related to local specificities. For instance, by merging data from the World Bank's Living Standards Measurement Study-Integrated Surveys on Agriculture (LSMS-ISA) with satellite or aerial imagery, policy and decision makers can obtain a more accurate view of the situation and issues that face different types of food producers in the world.[116] Private companies can propose new services to farmers to insure their production against weather risks.[117]

Better management of fields is especially crucial for "marginal landscapes" where food production is limited due to harsh weather conditions and limited water availability.[118] Israel offers a good case study of precision agriculture, or the use of technology to improve the efficiency of food production.[119] Precision agriculture encompasses first gathering big data from a large range of sources including sensors and satellite imagery, and then analyzing them to formulate precise recommendations about when and where to plant, how much water to use and when, what nutriments to add and when, what type of crop to plant, which land to use or to let lie fallow, etc. For instance, CropX is a software company that developed an advanced technology that provides this type of information to farmers, but also automatically controls the irrigation system based on a number of sensors placed in a field and connected to a GPS-enabled smartphone.[120] Taranis is another example of a big data application that collects data from sensors and satellites to provide advice on diseases and pest prevention. All information is then available on a digital dashboard to help farmers make decisions.[121] Tevatronic provides another type of autonomous irrigation system. Its innovation consists of wireless-connected sensors that collect data from the soil, but also measure the levels of stress of each plant to determine when to start and stop watering. It increases the productivity of crops up to 31% and allows savings on water and fertilizer of up to 75%.[122] Trellis collects big data from a large number of farmers, and then utilizes big data analytics combined with artificial intelligence to provide cloud-based concrete best-practice recommendations to improve levels of productivity and sustainability.[123] Hence, big data can improve the efficiency of food production, reduce waste, and reach SDG Goal 12, "Ensure sustainable consumption and production patterns."[124]

However, big data is not without critics. These recent technologies are sometimes based on data that are collected unintentionally, which raises questions of privacy and confidentiality. However, in the fields related to global environmental governance, this issue is limited. Nevertheless, two important challenges must be addressed: shareable data and biases. First, these large data sets are not always based on a common standard, which reduces their impact and the opportunities for innovation and nature conservation. Second, data collected by citizens can be inherently and unintentionally biased, which leads researchers to focus more attention on data integrity. Indeed, due to their intrinsic features, big data require new non-human instruments and techniques in order to be handled properly and to permit scientists to make sense out of them.[125] Among these new techniques, machine learning and algorithms help to distinguish patterns, trends, relationships, and dependencies, and to develop prediction models. However, in the case of volunteered and automated sources, the data collected might not be random, and therefore biases must be carefully considered in order to ensure the validity of the data.[126]

This chapter has aimed at illustrating how big data are used by civil society to enhance its role as knowledge broker on the international stage. It is not the only actor using this technology, as mentioned previously. And the list of applications of big data by civil society and other actors is also quite large and more varied than illustrated here. However, the choice was made to focus on one of the roles of civil society that will benefit the most from big data. Consequently to being a knowledge broker, civil society is also enhancing its role as advocate. More data and better analysis of the environment and human activities will allow it to influence more effectively the decision-making processes of global environmental governance thanks to websites[127] and social media platforms,[128] as discussed in previous chapters.

Notes

1 Phil, S., 2014. Potholes and big data: Crowdsourcing our way to better government. *Wired*. www.wired.com/insights/2014/03/potholes-big-data-crowdsourcing-way-better-government/

2 Seon-Cheol, Y., Dong-Bin, Shin, Jong-Wook, A., 2016. A study on concepts and utilization of geo-spatial big data in South Korea. *KSCE Journal of Civil Engineering*, 20(7), p. 2893.

3 Savage, N., 2018. Big data goes green: The global proliferation of Earth and environmental datasets opens new avenues for discovery. *Nature*, 558(19). doi:10.1038/d41586-018-05484-4

4 Richards, N.M., King, J.H., 2014. Big data ethics. *Wake Forest Law Review*, 49, p. 393.

5 Provost, F., Fawcett, T., 2013. Data science and its relationship to big data and data-driven decision making. *Big Data*, 1(1), p. 54.

6 Hellweg, E., 2012. The first big data election. *Harvard Business Review*. https://hbr.org/2012/11/2012-the-first-big-data-electi

7 Ibid.

8 Oracle, 2018. What is big data? *Oracle*. www.oracle.com/big-data/guide/what-is-big-data.html

9 Savage, N., 2018. *Op Cit.*

10 White, R.L., Sutton, A.E., Salguero-Gomez, R., Bray, T.C., Campbell, H., Cieraadn, E., Geekiyanage, N., et al., 2015. The next generation of action ecology: Novel approaches towards global ecological research. *Ecosphere*, 6(8), p. 134.

11 Lehrer, J., 2010. A physicist solves the city. *New York Times Magazine*. www.nytimes.com/2010/12/19/magazine/19Urban_West-t.html

12 Ford, J.D., Tilleard, S.E., Berrang-Ford, L., Araos, M., Biesbroek, R., Lesnikowski, A.C., MacDonald, G.K., et al., 2016. Big data has big potential for applications to climate change adaptation. *Proceedings of the National Academy of Sciences, USA*, 113(39), pp. 10729–10732.

13 Death, R.G., 2015. An environmental crisis: Science has failed; let us send in the machines. *Wiley Interdisciplinary Reviews: Water*, 2(6), p. 595.

14 Fleming, L.E., Haines, A., Golding, B., Kessel, A., Cichowska, A., Sabel, C.E., Depledge, M.H., et al., 2014. Big data has big potential for applications to climate change adaptation. *PNAS*, 113(39), p. 10731.

15 Dafforn, K.A., Johnston, E.L., Ferguson, A., Humphrey, C.L., Monk, W., Nichols, S.J., Simpson, S.L., Tulbure, M.G., Baird, D.J., 2016. Big data opportunities and challenges for assessing multiple stressors across scales in aquatic ecosystems. *Marine and Freshwater Research*, 67(4), p. 408.

16 Salmond, J.A., Tadaki, M., Dickson, M., 2017. *Op Cit.* p. 52.

17 Levin, N., Kark, S., Crandall, D., 2015. Where have all the people gone? Enhancing global conservation using night lights and social media. *Ecological Applications*, 25(8), p. 2153.

18 Steinle, S., Reis, S., Sabel., C.E., 2013. Quantifying human exposure to air pollution – moving from static monitoring to spatio-temporally resolved personal exposure assessment. *Science of the Total Environment*, 443, p. 188.

19 Dunkel, A., 2015. Visualizing the perceived environment using crowdsourced photo geodata. *Landscape and Urban Planning*, 142, p. 179.

20 Salmond, J.A., Tadaki, M., Dickson, M., 2017. *Op Cit.* p. 54.

21 Hsu, L., Martin, R.L., McElroy, B., Litwin-Miller, K., Kim, W., 2015. Data management, sharing, and reuse in experimental geomorphology: Challenges, strategies, and scientific opportunities. *Geomorphology*, 244, p. 180.

22 Salmond, J.A., Tadaki, M., Dickson, M., 2017. Changing priorities in physical geography: Introduction to the Special Issue. *Canadian Geographer / Le Géographe canadien*, 61(1), p. 54.

23 Editorial, 2008. Community cleverness required. *Nature*. www.nature.com/articles/455001a

24 Cox, M., Ellsworth, D., 1997. Application-controlled demand paging for out-of-core visualization. *Proceedings of the 8th conference on visualization*, Phoenix, AZ, p. 4.

25 Lokers, R., Knapen, R., Janssen, S., van Randen, Y., Jansen, J., 2016. Analysis of big data technologies for use in agro-environmental science. *Environmental Modelling & Software*, 84, p. 495.

26 Marr, B., 2015. Why only one of the 5 vs of big data really matters. *IBM Big Data & Analytics Hub*. www.ibmbigdatahub.com/blog/whyonly-one-5-vs-big-data-really-matters.

27 Bunn, J., 2012. How big is a petabyte, exabyte, zettabyte, or a yottabyte? *Highscalability*. http://highscalability.com/blog/2012/9/11/how-big-is-a-petabyte-exabyte-zettabyte-or-a-yottabyte.html

28 Levin, N., Kark, S., Crandall, D., 2015. Where have all the people gone? Enhancing global conservation using night lights and social media. *Ecological Applications*, 25(8), p. 2154.

29 Mishra, N., Lin, C-C., Chang, H-T., 2014. A cognitive oriented framework for IoT big-data management prospective. *2014 IEEE International Conference on Communication Problem-Solving (ICCP)*, Beijing.

30 RFID stands for radio-frequency identification.

31 Sarma, S., Brock, D.L., Ashton, K., 2000. The networked physical world. *Auto-ID Center White Paper MIT-AUTOID-WH-001*, Cambridge, MA.

32 Boos, D., Guenter, H., Grote, G., Kinder, K., 2013. Controllable accountabilities: The internet of things and its challenges for organisations. *Behaviour & Information Technology*, 32(5), p. 449.

33 Atzori, L., Iera, A., Morabito, G., 2010. The internet of things: A survey. *Computer Networks*, 54(15), p. 2.

34 Mayer-Schonberger, V., Cukier, K., 2013. Big data: A revolution that will transform how we live, work, and think. *American Journal of Epidemiology*, 179(9), p. 1143.

35 Salmond, J.A., Tadaki, M., Dickson, M., 2017. *Op Cit.*, p. 55.

36 Kitchin, R., 2013. Big data and human geography: Opportunities, challenges and risks. *Dialogues in Human Geography*, 3(3), p. 262.

37 Ibid.

38 Lokers, R., Knapen, R., Janssen, S., van Randen, Y., Jansen, J., 2016. *Op Cit.*, p. 496.

39 Ibid.

40 Salmond, J.A., Tadaki, M., Dickson, M., 2017. *Op Cit.*, p. 55.

41 Gura, T., 2013. Citizen science: Amateur experts. *Nature*, 496, p. 259.

42 Mason, M., 2017. Big data: Explaining its uses to environmental sciences. *Environmental Science*. www.environmentalscience.org/data-science-big-data

43 Chen, H., Chiang, R., Storey, V., 2012. Business intelligence and analytics: From big data to big impact. *MIS Quarterly Special Issue: Business Intelligence Research*, 36(4), p. 1166.

44 Hopkins, J., Hawking, P., 2018. Big data analytics and iot in logistics: A case study. *The International Journal of Logistics Management*, 29(2), p. 576. https://doi.org/10.1108/IJLM-05-2017-0109

45 Simon, P., 2014. Potholes and big data: Crowdsourcing our way to better government. *Wired*. www.wired.com/insights/2014/03/potholes-big-data-crowdsourcing-way-better-government/

46 Bertot, J., Estevez, E., Janowski, T., 2016. Editorial, universal and contextualized public services: Digital public service innovation framework. *Government Information Quarterly*, 33, p. 218.

47 Cool Green Science. https://blog.nature.org/science/2014/06/03/citizen-science-fieldscope-maps-water/

48 National Geographics. www.nationalgeographic.org/education/programs/fieldscope/?ar_a=1

49 Chesapeake Bay Watershed Project. http://chesapeake.fieldscope.org

50 Frog Watch. www.aza.org/frogwatch

51 Litterati. www.litterati.org

52 Bird Returns. http://birdreturns.org/home-page/about/about/

53 Raw data, 2018. Big data nature. *Stanford University Podcast.* http://worldview.stanford.edu/raw-data/episode-7-big-data-nature

54 Ibid.

55 Feeder Watch. https://feederwatch.org/about/project-overview/

56 Ibid.

57 NestWatch. https://nestwatch.org/about/overview/

58 GBBC. http://gbbc.birdcount.org/about/

59 African Wildlife Poinsoning. www.africanwildlifepoisoning.org

60 African Wildlife Poinsoning (map). www.tgpcloud.org/wildlife/

61 Conserve.io. http://conserve.io/our-work/

62 Whale Alert. www.whalealert.org

63 Save the Manatee. www.savethemanatee.org

64 SharkNet. https://hopkinsmarinestation.stanford.edu

65 Bumble Bee Watch. www.bumblebeewatch.org/about/

66 Xprize. *Op Cit.*

67 Function, 2014. Case Study. *HP Earth Insights.* www.aiga.org/case-study-hp-earth-insights

68 See Conservation International. www.conservation.org/partners/Pages/Hewlett-Packard-Enterprise.aspx

69 GEC. www.greatelephantcensus.com/the-census/

70 GEC, 2016. The great elephant census: Census results summary. *Paul G. Allen Project,* p. 1. www.greatelephantcensus.com/final-report

71 Sztutman, M., 2014. IBM, big data and saving the Amazon (the Real Amazon). *Nature.* https://blog.nature.org/conservancy/2014/09/17/ibm-big-data-and-the-amazon-the-real-amazon/

72 High Carbon Stock Approach. http://highcarbonstock.org

73 Kaye, L., 2016. Greenpeace taps big data and collaboration to halt deforestation. *Triple Pundit.* www.triplepundit.com/2016/09/business-forests-high-carbon-stock-approach/

74 Intel, 2014. Tiny chip helping save huge Rhinos. *Intel Free Press.* https://newsroom.intel.com/editorials/intel-galileo-quark-ankle-collar-bracelet-save-endangered-rhinos/

75 Goal 15. *Protect, restore and promote sustainable use of terrestrial ecosystems, sustainably manage forests, combat desertification, and halt and reverse land degradation and halt biodiversity loss.* See more https://unstats.un.org/sdgs/indicators/indicators-list/

76 LightHill Risk Network. https://lighthillrisknetwork.org/about-the-network/

77 University of Reading. www.reading.ac.uk/news-and-events/releases/PR604426.aspx

78 Ibid.

79 Madurika, H.K.G.M., Hemakumara, G.P.T.S., 2017. GIS based analysis for suitability location finding in the residential development areas of greater matara region. *International Journal of Scientific and Technology Research,* 4(8), p. 97.

80 Seon-Cheol, Y., Dong-Bin, S., Jong-Wook, A., 2016. A study on concepts and utilization of geo-spatial big data in South Korea. *KSCE Journal of Civil Engineering,* 20(7), p. 2893.

81 Rahman, S., DiLiping, L., Ul-Zannat, E., 2017. The role of big data in disaster management. *Proceedings conference: International conference on disaster risk mitigation,* Dhaka, Bangladesh, p. 22.

82 Bennett, V.L., Kershaw, P., Busswell, G., Hilton, R., O'Neill, A., 2013. CEMS: A new infrastructure for EO and climate science. *Proceedings: ESA living planet symposium 2013*, Edinburgh, UK, 9, p. 1.

83 Ibid.

84 Xprize, 2018. *AI solving sustainable development goals*. https://ai.xprize.org/AI-For-Good/sustainable-development-goals

85 Savage, N., 2018. *Op Cit.*

86 Ibid.

87 Ibid.

88 Ibid.

89 Ibid.

90 Rejcek, P., 2017. Big data is helping us see environmental problems in a whole new light. *SingularityHub*. https://singularityhub.com/2017/01/23/big-data-is-helping-us-see-environmental-problems-in-a-whole-new-light/

91 Moran, D., Kanemoto, K., 2017. Identifying species threat hotspots from global supply chains. *Nature Ecology & Evolution*, 1. doi:10.1038/s41559-016-0023

92 Sanders, R., 2014. Big data guides conservation efforts. *UC Berkeley*. http://news.berkeley.edu/2014/07/18/big-data-guides-conservation-efforts/

93 Xprize. *AI solving sustainable development goals*. https://ai.xprize.org/AI-For-Good/sustainable-development-goals

94 GTOPP. www.topp.org

95 Death, R.G., 2015. An environmental crisis: Science has failed; let us send in the machines. *Wiley Interdisciplinary Reviews: Water*, 2(6), p. 596.

96 Janssen, M., Konopnicki, D., Snowdon, J., Ojo, A., 2017. *Op Cit.*, p. 194.

97 Global Pulse Initiative. www.unglobalpulse.org/about-new

98 Ibid.

99 Giest, S., 2017. Big data for policymaking: Fad or fast track? *Policy Science*, 50, p. 367. doi:10.1007/s11077-017-9293-1

100 Manogaran, G., Lopez, D., 2018. Spatial cumulative sum algorithm with big data analytics for climate change detection. *Computers & Electrical Engineering*, 65, p. 207. https://doi.org/10.1016/j.compeleceng.2017.04.006

101 Apte, J.S., Messier, K.P., Gani, S., et al., 2017. High-resolution air pollution mapping with google street view cars: Exploiting big data. *Environmental Science and Technology*, 51(12). doi:10.1021/acs.est.7b00891

102 Atkins, J., Maroun, W., Atkins, B.C., Barone, E., 2018. From the big five to the big four? Exploring extinction accounting for the rhinoceros. *Accounting, Auditing & Accountability Journal*, 31(2), p. 687. doi:10.1108/AAAJ-12-2015-2320

103 Anderson, K., Ryan, B., Sonntag, W., Kavvada, A., Friedl, L., 2017. Earth observation in service of the 2030 Agenda for Sustainable Development. *Geo-Spatial Information Science*, 20(2), p. 2. doi:10.1080/10095020.2017.1333230

104 UN World Data Forum. https://undataforum.org/WorldDataForum/about/

105 Burchette, C., 2016. Filling the gaps in environmental science with big data. *Environmental Protection Agency's Blog*. https://blog.epa.gov/blog/2016/10/filling-the-gaps-in-environmental-science-with-big-data/

106 Ibid.

107 EPA StreamCat. www.epa.gov/national-aquatic-resource-surveys/streamcat

108 EPA EnviroAtlas. www.epa.gov/enviroatlas

109 EPA Web-based Interspecies Correlation Estimation (Web-ICE). www3.epa.gov/webice/

110 Burchette, C., 2016., *Op Cit.*

111 USEPA Environmental Quality Index (EQI) – air, water, land, built, and sociodemographic domains transformed variables dataset as input for the USEPA EQI, by County for the United States. https://catalog.data.gov/dataset/usepa-environmental-quality-index-eqi-air-water-land-built-and-sociodemographic-domains-transf

112 NASA Goddard introduces the NASA center for climate simulation. www.nasa.gov/centers/goddard/news/releases/2010/10-051.html
113 EOSDIS Program. https://earthdata.nasa.gov/about
114 Savage, N., 2018. *Op Cit.*
115 Woodard, J., et al., 2017. Using ICT for remote sensing, crowdsourcing and big data to unlock the potential of agricultural data. *ICT in Agriculture*, p. 402. http://ilsirf.org/wp-content/uploads/sites/5/2017/08/2017 _WorldBank_Chapter_15.pdf
116 Ibid.
117 Ibid.
118 Savage, N., 2018. Big data goes green: The global proliferation of Earth and environmental datasets opens new avenues for discovery. *Nature*, 558(19). doi:10.1038/d41586-018-05484-4
119 Klein Leichman, A., 2017. 5 Israeli precision-ag technologies making farms smarter. *Israel21c*.www.israel21c.org/5-israeli-precision-ag-technologies-making-farms-smarter/
120 CropX. www.cropx.com
121 Taranis. www.taranis.ag
122 Tevatronic. http://tevatronic.net
123 Trellis. www.trellis.ag
124 Ibid.
125 Death, R.G., 2015. An environmental crisis: Science has failed; let us send in the machines. *Wiley Interdisciplinary Reviews: Water*, 2(6), pp. 595–600.
126 Dickinson, J.L., Zuckerberg, B., Bonter, D.N., 2010. Citizen science as an ecological research tool: Challenges and benefits. *Annual Review of Ecology, Evolution, and Systematics*, 41, p. 166.
127 See Chapter 2 of this book for an analysis of the use of websites by CSOs.
128 See Chapter 3 of this book for an analysis of the use of social media platforms by CSOs.

7 Artificial intelligence and environmental civil society organizations

1. Introduction

In an interview with the BBC in December 2014, Professor Stephen Hawking warned the world against the dangers of full artificial intelligence (AI) that "could spell the end of the human race."[1] He further argued that "It would take off on its own, and re-design itself at an ever increasing rate,"[2] in which case "Humans, who are limited by slow biological evolution, couldn't compete, and would be superseded."[3] He is not the only expert to suggest a careful approach to artificial intelligence. Elon Musk, the founder of Tesla and SpaceX, shares this gloomy perspective, stating that "AI could become an immortal dictator from which we would never escape."[4] However, as much fear as it triggers, AI also inspires great expectations. Russian President Vladimir Putin explained that AI represents the future for all humankind and that "It comes with colossal opportunities, but also threats that are difficult to predict. Whoever becomes the leader in this sphere will become the ruler of the world."[5] Similarly, former Google CEO Eric Schmidt compares AI to space technology during the cold war, calling for a national security strategy on AI[6] as it is becoming a vital element for state power.

Beyond the arenas of politics, business, and academia, AI has become mainstream over the last few years, along with popular movies such as *The Terminator*, *The Matrix*, and *Ex Machina*. Indeed, AI is becoming more and more visible in our daily lives through voice-powered personal assistants (PAs) such as Google Assistant, which can book a table and chat with people on the phone, or self-driving cars.[7]

AI is a general-purpose technology, similar to electricity in the sense that its applications are endless, and it will affect society at large, much more than the invention of a submarine for instance.[8] The humanoid robot Sophia, developed by Hanson Robotics, has given a face to AI and a glimpse of the future interaction between robots and humans. Its speech at the United Nations (UN),[9] but also on various news channels, surprised many, and even scared some journalists, when it demonstrated the ability to laugh, to tell a joke, and to offer a romantic recommendation.[10]

Max Tegmark, in this book entitled *Life 3.0*,[11] associates the emergence of general artificial intelligence to a third phase in evolutionary history. For about four billion years, biology restricted the evolution of the (human) body and brain. Through learning and culture, humans have developed the capacity of the latter. AI represents the third phase, where body and brain can be re-engineered and expanded without biological limitations.[12] As *The Economist* argued: "The implications of introducing a second intelligent species onto Earth are far-reaching enough to deserve hard thinking."[13] In 2015, AlphaGo® software won against a professional Go player without handicaps on a full-sized 19×19 board.[14] This victory was possible only by combining strategy and calculation skills. AlphaGo developed its strategy skills by playing against itself millions of times. This leads us to think that AI could develop superior negotiation skills, which could be applied to contract negotiations or hostage situations.[15] Most concerns are not related to current AI applications, but rather to its future developments, questioning whether they will lead to technological singularity, when human beings will no longer be the most intelligent species on the planet.[16]

Artificial intelligence holds many promises for accelerating the achievement of the Sustainable Development Goals (SDGs), in particular in predicting the effects of climate change and the path of storms, monitoring the spread of diseases, helping understand human decision-making processes in terms of energy consumption and protection of the environment, or identifying areas most in need of attention. Various research projects have used the power of AI to predict poverty in Nigeria, Uganda, Tanzania, Rwanda, and Malawi.[17]

AI is increasingly taking a leading role in making sense of the large volumes of data recorded by satellites, drones, smartphones, and sensors throughout the planet. This new technology already allows us to better understand our planet, from its urban communities to its most remote and untouched areas, from high mountain peaks to deep oceans. AI can support global environmental governance in multiple ways. This chapter focuses on the use of the technology by civil society. However, some illustrations of its use by local public authorities and international organizations will be mentioned in the final remarks. Civil society organizations uses AI in multiple ways, but mainly for two purposes: providing better environmental services, and producing new scientific knowledge. This chapter will first discuss the concept of artificial intelligence, and then examine two applications, namely implementing field projects and producing knowledge.

2. Artificial intelligence explained

John McCarthy was the first to coin the term "artificial intelligence" in a proposal for the famous 1956 Dartmouth Conference, which started AI as a field of research.[18] Although AI encompasses a broad range of activities, it involves machines that accomplish duties that are specific to human intelligence, and includes tasks such as planning, understanding language, learning, and problem solving.[19]

The term "artificial intelligence" is somewhat misleading. Indeed, the term "artificial" implies that it is unreal, while "intelligence" refers to the human capacity to manage complex tasks and operations. In fact, AI is real, and its impact on many fields is already substantial. Furthermore, the use of AI today is not restricted to tasks that would be considered as *intelligent*, but includes rather simplistic ones such as recognizing objects in pictures.[20]

Since its first developments, research in AI has been divided into two separate camps. On the one hand, symbolic AI, introduced by Newell & Simon in 1976, refers to a system based on rules and knowledge, where computers manipulate symbols rather than letters or numbers. This approach solves problems through their symbolic representation. Such symbols can be arranged in lists, hierarchies, and networks, allowing the detection of relationships between them; these are called production rules, and they are similar to If-Then statements. A form of symbolic AI is the expert system, which uses production rules to make deductions and decisions. Also called the Good-Old-Fashioned Artificial Intelligence (GOFAI), symbolic AI is defined by an exclusive emphasis on symbolic reasoning and logic.[21] For instance, IBM's Deep Blue, the AI technology that won a chess tournament against Garry Kasparov in 1997, used this approach.[22]

On the other hand, non-symbolic AI, also called computational or connectionist AI, focuses on calculation according to rules that have been pre-established. This approach solves problems through calculation, using systems of large networks of simple numerical processors that communicate with each other and perform tasks in parallel.[23] Connectionist AI is intended to imitate how a human brain functions and in particular its complex network of neurons.[24] Some current applications of connectionist AI are well known: Google's automatic transition system (which looks for patterns), Facebook's face recognition algorithm, and AI enabling self-driving cars.

AI encompasses a wide range of applications, and it can be classified in three categories: Artificial Narrow Intelligence (ANI), Artificial General Intelligence (AGI), and Artificial Super Intelligence (ASI).[25]

The first category of AI corresponds to the most visible and current applications of AI. ANIs have exceptional calculation capabilities, but they are restricted to a specific area, such as IBM's Deep Blue® that beat Gary Kasparov at chess in 1997. More recent applications include Apple's Siri®, search engines, algorithms of social media platforms, web cookies that identify users online, digital advertising, data miners, self-driving cars, traffic control software, etc. Although highly efficient for their dedicated tasks, ANIs cannot embark on other operations that were not pre-determined by the creators of the ANI. Millions of YouTube videos are captured in a large number of languages thanks to the ANIs that efficiently convert sound into text, and then translate it.[26]

Contrary to the case with humans, performing complex calculation is simple and fast and does not require much effort for AIs. However, tasks that are based on perception and anticipation, and that require assessing and processing random situations, for instance voice and image recognition,[27] are difficult. In other words, "AI has by now succeeded in doing essentially everything that requires 'thinking' but has failed to do most of what people and animals do 'without thinking', that, somehow, is much harder."[28]

This is in fact what the second category of AI aims to overcome. AGIs correspond to the capacity of the human brain to process information, solve complex problems, perform tasks, and make decisions in many different fields. AGIs could on one hand help human beings grow beyond their biological legacy and increase the diversity of the human experience, while on the other hand develop the capacity to independently control their own judgments, concerns, feelings, forces, vulnerabilities, and biases.[29] This second category of AIs is for the future, and rare are the scientists who agree on when it will become available. Some believe it will become reality already in 2030, whereas others believe it will simply never happen.

The third category relates to AI that outperforms the human brain in all tasks including scientific creativity, general wisdom, and social skills.[30] After reaching AGI, ASI will appear quickly[31] as AGI will develop its capacity exponentially according to an "intelligence explosion" or "technological singularity" phenomenon.[32] This is the stage of AI development that society and governments fear the most, since it illustrates how human beings can be superseded by machines.

Algorithms are at the heart of AI. Similar to a lever for a watch, algorithms enable AI to perform calculations. Algorithms correspond to a set of pre-defined instructions that indicate what to do in specific situations. In fact, AI often uses a number of interrelated algorithms that work hand in hand. The simplest algorithm is the one that provides the instruction to the computer to activate the switches (or transistors) that make up a computer: if the transistor is on, it is equal to 1; if it is off, it is equal to 0.[33]

The instructions involve the use of several transistors that are linked to each other. For instance, if transistors B and C are off, then transistor A must turn on. This combination of actions based on the conditions "IF," "AND," and "OR" represents the basis of all algorithms. This allows computer algorithms to receive data, process them, and then produce a specific outcome such as: "IF" circumstance A "AND" circumstance B, "OR" circumstance C, then prescribe "X medication."[34] Another feature of algorithms is that instructions must be accurate and non-ambiguous. In other words, a home-made recipe for a chocolate cake could not be reproduced exactly by a computer every time, since how much is a spoonful of sugar exactly? Therefore, such a recipe is not an algorithm.[35]

Machine learning is when an algorithm develops the capacity to create new algorithms. Called "learners," these algorithms function differently. As input, they receive large amounts of data (such as millions of pictures of cats) and the desired outcome[36] (such as the capacity to identify all types of cats in pictures). As output, the learner analyzes all pictures and creates a set of instructions (a new algorithm) that can describe and therefore identify all kinds of cats. In that sense, big data are closely linked to machine learning, since they provide the input for learners to create new algorithms.

As part of the connectionist AI approach, machine learning is the ability of a machine to learn by itself. So instead of hard coding software routines with specific instructions to accomplish a particular task, machine learning trains a program to learn how to act in situations for which it was not programed. The training refers to providing large amounts of data to the program so that it can learn and extract meaning from those data.[37] A good example of machine learning is image recognition. A computer program is fed with large numbers of images from social media platforms for instance. Each picture is identified by users as a cat, car, or specific person. The program can then learn to identify the characteristics of each object or person, which will enable it to later recognize them by itself even when they are not tagged by users. In a way, the program has learned to define what a cat, car, or specific person looks like.[38] This is a characteristic of AI, since previously, the computer had to be instructed specifically about all possible forms of an object or a person to be able to perform an identification.

Machine learning consists of many different approaches such as decision tree learning, inductive logic programming, clustering, reinforcement learning, Bayesian networks, and deep learning. The latter is probably the most well-known to the general public. It involves Artificial Neural Networks (ANNs), which consist of interconnected "neurons" with multiple layers. Each layer is dedicated to a specific aspect of the learning process: the edges of a photo for instance.

Prior to discussing AI's concrete applications, some of its main technical applications will be considered. First, AI has the capability of environmental perception by collecting and transforming pictures, sounds, texts, articles, and large amounts of data. Second, AI has the ability to process that information and make sense of it; it can identify patterns and schemes, similar to human interpretation, although it cannot draw true meaning out of its analysis. Third, AI can make decisions and act upon them based on the collection of data from its environment and its analysis. Lastly, AI can learn through the data collection and analysis, which enables it to constantly improve its capacity to succeed in performing specific tasks.[39]

Based on these abilities, AI planning corresponds to the capacity of AI to choose a sequence of actions that should (according to its representation of the model of the world) have a specific and pre-defined impact on the real world. Some complex forms of AI planning deal with uncertainty, involving multiple agents and an environment that cannot be fully observed, which means that external conditions can be only partially assessed but not with certainty.[40]

Machine perception refers to this ability of AI to combine pieces of information collected from numerous sensors in the physical world, and interpret them to create one single image of the world. It functions similarly to the human eye, which collects information from the outside world; this information is then combined in the brain to form one single image through a principle called sensor fusion. Some applications of machine perception include handwriting recognition, image processing, and document analysis.[41] The capacity of AI to understand human language is called natural language processing (NLP). Among its multiple applications are text mining, responding to enquiries, and language translation.

Many robots use AI to assess their environment and perform specific tasks. Moravec's paradox states that skills that require thinking or reasoning, in other words high computational capacity, are easy for AI and difficult for humans. On the other hand, unconscious skills, such as recognizing a face or learning how to move in an environment, are easy for humans but extremely difficult for AI. One reason is that unconscious skills were developed over millions of years of evolution. And while they seem easy for humans, they are nevertheless highly complex.[42] This implies that motion and coordination in a multi-agent environment and affective computing, which describes the ability of AI to assess, interpret, and simulate human affects,[43] are more difficult for AI than solving a mathematical problem.

AI can help solve a large number of contemporary challenges. As discussed in this section, AI[44] is at an early stage. The scope of its impact is yet to be determined. Nevertheless, it is already possible to envisage some of its current applications in the field of global environmental governance, and in particular, how civil society is embracing this technology to improve its roles as service provider and knowledge broker. The Sustainable Development Goals, by offering a global agenda for development, provide an ideal framework to envisage how AI can empower global and local development. As shown in this chapter, AI can support many SDGs. The following section will examine some current and emerging AI applications.

3. Implementing field projects

Civil society participates fully in the implementation of field projects throughout the globe. Either independently or through the delegation of an international organization or a state, civil society has developed field expertise and know-how in terms of protection of the environment. The environmental services this group of actors provides are numerous, starting from protecting endangered species and ecosystems to tracking poaching crimes and overfishing. They also support other actors in educating and training parts of the population to better manage natural resources.

On land, biodiversity loss seems unstoppable. Thanks to image recognition technologies, game theory, and AI analysis capacities, animals in the wild are better monitored and protected with real-life data and accurate prediction models.[45] Poaching incidents can be reduced, pests and viruses can be detected at an early stage, and migrations can be protected by creating temporary wetlands in collaboration with farmers.[46]

Forests can provide one-third of the solutions to climate change and are essential for the conservation of biodiversity.[47] However, they are under multiple threats: deforestation, but also degradation with a decline in density and underbrush, which leads them to be more vulnerable to fire for instance. Despite the current advances in techniques to assess restoration efforts remotely, the lack of real-time data continues to make it difficult to evaluate whether or not the concessions and payments made to the states and enterprises in charge of forest protection are effective.[48] In that context, satellite and aerial mapping, along with the analysis capacity of AI, can provide in real time substantive and accurate information about the degradation of underbrush, thinning of upper-story trees, evolution of water levels, and the increase of new trees in designated areas.[49] Hence, this illustrates how AI can help achieve SDG Goal 15, which seeks to "Protect, restore and promote sustainable use of terrestrial ecosystems, sustainably manage forests, combat desertification, and halt and reverse land degradation and halt biodiversity loss."[50]

The Chesapeake Conservancy, in collaboration with Microsoft, has tackled the need to more accurately understand changes in complex ecosystems such as watersheds. The initiative aims at creating high-resolution land cover maps to support scientists and decision makers in protecting nature in the United States. Data are collected from a range of sources such as drones, airplanes, and satellites and stored on a cloud system. AI is then used to analyze these large volumes of data to produce up-to-date land cover maps offering more than 900 times the information that was available previously, allowing for precision conservation measures.[51]

When considering wildlife protection, one of the main issues that comes to mind is poaching. For instance, an average of 96 African elephants are killed by poachers every day, leading scientists to believe that this animal could disappear entirely in the coming decades if nothing is done.[52] Traditional attempts to stop poaching are based on rangers who regularly patrol large geographic areas. For rangers to know where poachers will be and how they will act is essential. In that respect, AI can help their work substantially.

With an elephant killed every 14 minutes, and two rhinos every day, these endangered species are at high risk. The Lindbergh Foundation collaborated with the tech company Neurala to create the Air Shepherd program, a combination of surveillance drones and artificial intelligence to protect elephant and rhino populations in Africa.[53] First, the use of drones equipped with infrared sensors helps spot poachers at night and sends their location to rangers who can then stop them before they reach the animals. Since the drones produce terabytes of video, AI assists human analysts to identify animals, vehicles, and poachers. Furthermore, AI helps predict where poachers will most likely be in order to intercept them in advance.[54] Air Shepherd was deployed at Liwonde National Park and Nkhotakota Wildlife Reserve in Malawi, Ezemvelo KZN Wildlife in South Africa, and Hwange National Park in Zimbabwe. Furthermore, in collaboration with the World Wide Fund for Nature (WWF) and Google, Air Shepherd was extended to Hwange National Park to stop poachers from poisoning watering holes.[55]

The Protection Assistant for Wildlife Security (PAWS) applies AI to collect data from previous poaching incidents, but also from terrain specificities such as hills, mountains, rivers, lakes, and forests as well as known migration paths for certain animals.[56] AI processes the data to detect where the next poaching activities are most likely to occur, and can thus guide rangers by providing them with the best route to patrol large geographic areas. The objective is to understand poachers' strategies and behaviors, to make predictions of their future moves, but also to adapt protection strategies as the poachers change theirs.[57] Moreover, AI helps rangers identify where the animals will be.[58]

Thanks to a continuous feed of data originating in the field, AI and in particular machine learning always adapt the routes and become more and more precise.

PAWS also selects some routes randomly to prevent poachers from identifying any patrolling pattern. PAWS is based on an algorithm that was developed according to the Stackelberg security game theory. This model of real-world security scenarios helps one to understand how adversaries think, when they will most probably attack, and how best to allocate resources. The security forces first deploy a mix of actions to protect an area for instance, and the attackers surveil these actions and then respond in the form of an attack. This algorithm is used today by a wide range of security forces, including US Homeland Security, the Transportation Security Administration, and the Coast Guard, to predict where to best locate security forces to stop smuggling and terrorism.[59]

The green version of this security game theory, called Green Security Games, adds the specificity of wildlife conservation conditions by taking into consideration the fact that interaction between defenders and attackers takes place over multiple rounds. First, the data collected about previous poaching are analyzed by the AI application to understand how the attackers behave. Second, it generates the best route for rangers to patrol. Third, the rangers (defenders) use this route to patrol the area. Fourth, the attackers observe the strategy executed by the defenders and respond by attacking an animal. This generates new data, which is fed to PAWS, and the loop starts again.[60]

In 2014, PAWS was first tested in a national park in Uganda and in a protected area in Malaysia.[61] Specifically, the Ugandan Wildlife Authority (UWA) fed PAWS with 14 years of data, including GPS coordinates, corresponding to more than 125,000 observations of poaching, and stemming from animal sightings, traps, animal carcasses, and other evidence. Some shortcomings were identified; for instance, the specific topography of the area (high altitude) and patrol schedules needed to be taken into consideration in the algorithm calculation. Nevertheless, in both cases PAWS proved to be useful in guiding rangers to poaching hotspots in these areas, and succeeded in decreasing the number of human activity signs sighted per kilometer.[62]

PAWS provides a good example of collaboration between AI and human intelligence. In this case, and contrary to other examples in this chapter, AI technology does not replace human interventions, but rather empowers them. This is the case with the next generation of PAWS, which is called INTERCEPT. At Uganda's Queen Elizabeth National Park, INTERCEPT led to catching a poacher and discovering a dozen elephant traps before they could be activated.[63]

Another AI-based wildlife preservation initiative was led by Deakin University in Australia with the objective to substantially improve animal monitoring. To date, the institution in charge of park management has used cameras and sensors to track animal movements and population. However, weather conditions trigger cameras and sensors that take irrelevant photos, leading to a large number of pictures without any animals. These photos had to be verified one by one by humans, which limited their positive impact on park management strategies. Thanks to AI, and in particular to machine learning and pattern recognition, these photos can be verified by the algorithm, which can sort through thousands of photos and keep only the ones with animals. The machine-learning feature of the algorithm will also allow it to identify what breed of animal has been caught by the camera. This new initiative will dramatically improve the efficiency and effectiveness of park management in the state of Victoria, Australia.[64]

The BeeScanning app[65] is an AI initiative dedicated to beekeepers.[66] Honeybees suffer from various threats, and in particular Varroa destructor. These red mites feed on the bees by crawling on their bodies and sucking their blood, which can lead to the destruction of entire hives if not dealt with in time. Consequently, every year, millions of honeybee colonies disappear,[67] which can have disastrous effects on the global food supply. Therefore, the objective is to detect honeybee hives that are infected by these red parasites as early as possible in order to treat them. Moreover, BeeScanning can also help identify which bees are most resistant to the red mites, which will allow scientists to comprehend the Varroa-resistant strain of bees.[68] The beekeeper takes a photo of the hive with the BeeScanning app. The photo is uploaded into the database, and then AI analyzes the images and determines if the bees are infected or not. The image analysis technology is based on Artificial Neural Networks, which have the ability to improve their detection capabilities over time. A deep learning professional trains the ANN to identify the red mites in the data collected. Through a process of errors and corrections, the ANN learns how to identify the abstract features of Varroa destructor. It then becomes faster and more accurate than the human eye in identifying early signs of the red mites in a hive. BeeScanning has a performance error of about ±5% in lab environment conditions.[69] Moreover, the data collected about performance, location, and date can help scientists in developing innovative protection solutions.

In the developing world, access to accurate data about wealth distribution and poverty remains a challenge, with few censuses being conducted, and data rarely collected from the poorest areas, where not even births and economic activities are recorded. However, data are crucial to establishing effective development plans and assistance mechanisms. AI can help overcome this lack of data through the analysis of satellite imagery for instance, allowing poverty to be mapped by using satellites to detect the lack of lighting infrastructure.[70] By allowing the analysis of large volumes of accurate and up-to-date data, AI can support efficient resource allocation and help reach SDG Goal 1 entitled "End poverty in all its forms everywhere."[71] Furthermore, thanks to predicative analysis from data emanating from the field (low-cost sensors, wireless-connected) and from the air (satellites and drones), AI can improve arable land productivity and reduce the environmental impact of agriculture through precision agriculture or data-driven agriculture.[72]

Indeed, SDG Goal 2 strives to "End hunger, achieve food security and improved nutrition and promote sustainable agriculture."[73] With two billion individuals relying on local agriculture to provide food, an increase in agricultural productivity can have a substantial impact on poverty reduction. However, climate change leads to rapidly changing environmental conditions, and a higher level of hazards, which makes planning for crop planting and harvesting even more of a challenge for local communities. In that context, AI can support agriculture by providing additional information to farmers to help them mitigate risks and increase their productivity, which leads to decreasing poverty overall.[74]

An AI project supported by Microsoft is focusing on farming in collaboration with the government of India, the Gates Foundation, BASF, Bayer, Land O'Lakes, and Mahindra. Similar to big data enabling precise farming, AI supports data-driven agriculture. In a world where the growing population puts additional pressure on natural resources and arable land, where water levels are decreasing and desertification is on the rise, data can be a solution to increase food production and respond to the additional need for food in the coming decades. FarmBeats is a combination of multiple technologies, including machine-learning algorithms, sensors, and drones, that provide insights to farmers so they can better maximize harvesting productivity, reduce costs, and decrease the environmental impact of agriculture, such as the amount of water used.[75] However, data-driven agriculture is a complex task to implement, since in many cases, there is no electricity or internet where the data must be collected. FarmBeats offers a solution to this issue. Low-cost sensors are not connected through Wi-Fi but through TV White Spaces radios, which allow the collection of data even without internet or electricity in the field. Machine-learning algorithms combine the data collected in the field with aerial imagery, best practices, and prediction data to produce almost real time and accurate actionable insights to local farmers.[76]

If AI can help civil society to improve its capacity to implement field projects and better protect the planet, and its most endangered species and ecosystems, the technology can also support the production of new knowledge that can then feed into policy- and decision-making processes on the international stage.

4. Producing knowledge

The AI capacity to analyze large volumes of data coming from satellite imagery is crucial to helping achieve the SDGs. Due to the interrelation and interdependency between the SDGs and the challenges that global environmental governance aims to tackle, the possibility to assess multiple criteria at the same time and on a global scale can support effective policy- and decision-making processes in today's world. The analysis of data stemming from weather conditions, animal population abundance and distribution, changes in ecosystems such as deforestation and desertification, and the spread of pests and viruses, among others, can help produce accurate prediction models based on multiple criteria and real-time conditions. Thanks to its predictive capacity, AI supports the development of accurate climate change models, and helps prepare for climate-related issues and disasters, in support of SDG Goal 13 to "Take urgent action to combat climate change and its impacts."[77]

Project Premonition is part of Microsoft's initiative entitled AI for Earth Project, which aims at providing AI capacity to solve some of the current major environmental challenges.[78] The objective of Project Premonition is to provide accurate data about biodiversity. Traditional data collection is time and resource intensive. Project Premonition innovates by utilizing mosquitoes as field scientists. The idea is to collect and analyze the blood taken by these insects from other animals to gather important information about the biodiversity in a specific area. Project Premonition combines several technologies to achieve this objective: drones are used to identify mosquito hotspots, robotic traps collect them, and cloud-scale genomics and machine-learning algorithms analyze the blood to identify each animal they bit. This innovative collaboration of technology, human expertise, and nature allows scientists and conservationists to gather more accurate and real-time information about the state of nature in a specific ecosystem.[79]

SDG Goal 14 entitled "Conserve and sustainably use the oceans, seas and marine resources" can be vastly enhanced thanks to the AI capacity to analyze large volumes of data stemming from wildlife abundance and distribution, allowing for a better understanding of migration, threats, and adaptation to current challenges such as overfishing, climate change, and ocean acidification, to name only a few.[80]

Overfishing, plastic pollution, and global warming are some of the numerous challenges that threaten marine wildlife. Coral reefs are particularly vulnerable to these changes, resulting in a 40% loss of corals worldwide in the last 30 years.[81] The XL Catlin Seaview Survey aims to track and communicate these transformations globally. It is a vast network of public and private scientific organizations including the Global Change Institute (GCI) of the University of Queensland, Google, Underwater Earth, Fourth Element, Lady Elliot Island, Mission-Blue, World Resources Institute (WRI), International Union for Conservation of Nature (IUCN), National Oceanic and Atmospheric Administration (NOAA), Scripps Institution of Oceanography (UC San Diego), and Living Oceans Foundation.[82]

The XL Catlin Seaview Survey started in 2012 and is developing a record of the coral reefs in the world in high-resolution, 360-degree panoramic vision to better monitor coral abundance, health, structure, and biodiversity globally, thereby supporting policy and decision makers in assessing and protecting coral reefs.[83] Monitoring produces large volumes of images, which are not possible to sort manually. The XL Catlin Seaview Survey utilizes an Artificial Neural Networking algorithm to identify what is on the photos and classify all the data.[84] About 81% of the time, the features in coral photos that are identified by the algorithm coincide with those identified by the human eyes of scientists; this percentage is in fact similar to that of two experts who compare their analyses of a coral photo. This combination of semi-automated data collection and monitoring opens the door to a new era of under-sea mapping of reefs in the world, which will allow scientists not only to have more accurate and up-to-date data about marine life, but also to dedicate more resources to research rather than to data collection and processing.[85]

CoralNet is an initiative from the University of California San Diego, which aims at reducing the annotation bottleneck. Indeed, while large amounts of data about marine life are collected, their analysis remains slow when each photo must be inspected manually. CoralNet uses an image analysis technology, which allows 50–100% automation.[86] HyperDiver is a project that monitors underwater wildlife with high precision, and then feeds these data into a machine-learning software similar to CoralNet, allowing the identification with high accuracy of proper taxonomic types including sponges, macroalgae, and seagrass.[87]

FishFace is a project that uses machine learning to identify wildlife diversity in deep oceans. The idea behind this project, which won the 2016 Google Impact Challenge, is to provide accurate data in order to better protect wildlife. Indeed, 90% of fisheries in the world are overexploited and lack stock assessment data.[88] To address overfishing requires having access to accurate and updated data about fish population distribution and abundance. This requires significant resources when done traditionally, which explains the lack of data in the developing world. The Nature Conservancy has developed innovative instruments to assess fish stocks affordably and in collaboration with fishers, government, and industry in Indonesia. FishFace uses machine learning to identify fish species. Data are collected through recycled smartphones used by fishers on their boats in the region. Thanks to image recognition technology, FishFace can quickly and accurately identify fish species and length from the photos taken by fishers, which will not only provide precise and up-to-date data about fish stock in the region, but also enable the fish to be sorted before arriving at the processing facility. The objective is to develop an affordable and efficient framework for fisheries assessment and management in the developing world.[89]

AI can further support the fight against plastic pollution. Although plastic production started in the 1950s with exponential growth since the 1980s, only a minority of the population was aware of the problem before the first decade of the 21st century, when social media helped raise awareness about the scale of plastic pollution in the ocean.[90] Thanks to the use of drones to map hotspots of plastic wastes, and AI to help identify plastics in the thousands of aerial photos taken, the NGO Plastic Tide[91] provides an accurate, open-source map of the most polluted coastlines. This then allows governments to better target clean-up measures, and better assess the effectiveness of policies to reduce plastic waste such as a ban on plastic bags for instance. Originally started in the United Kingdom, Plastic Tide aims to develop its operations globally.[92]

Moreover, citizen science and artificial intelligence go hand in hand to protect wildlife. Thanks to the power of citizen conservationists who collect large volumes of data throughout the world, and to AI that can help analyze these big data, scientists now have access to a wealth of insights never foreseen. In this context, AI allows for innovative forms of collaboration between academia, governments, civil society, and the private sector, helping reach SDG Goal 17 entitled "Revitalize the global partnership for sustainable development."[93] Furthermore, AI can support remote learning programs and improve participation and completion rates thanks to virtual mentors and responsive individualized learning programs. Remote education offers a wide range of courses, in particular related to nature conservation, which are accessible 24/7 around the globe for free or at a more affordable price than traditional onsite education, supporting SDG Goal 4, "Ensure inclusive and quality education for all and promote lifelong learning."[94]

Most wildlife research initiatives lack resources. Collecting and managing data are often not systematic and, in the long run, are often dedicated to a one-off analysis for a specifically funded research project. With limited global or regional data sharing between projects, this leads to a difficulty in reaching a critical mass of data for endangered species. In addition, traditional data collection strategies and manual data processing, such as verifying images manually, are resource and time intensive, leading to limited scales and to some delay between the observation phase and the recommendations. ICTs and in particular AI can help overcome these limitations in terms of scope, scale, repeatability, continuity, and return on investment[95] by providing highly efficient and less resource-intensive data collection, management, and analysis instruments.

iNaturalist is one of the most popular nature apps, which enables users to identify plants and animals in their surroundings; it was developed in a collaboration between the California Academy of Sciences and the National Geographic Society. iNaturalist has succeeded in creating a community of citizen conservationists all around the world who share their observations through this app, and receive information from a network of more than 750,000 scientists.[96] The idea of this initiative is to combine two objectives: to connect people to nature by enabling them to better understand nature around them, and to collect scientifically valuable biodiversity data. Once a photo is taken by a user in the field, AI along with computer vision models make a suggestion about the sighted species; this suggestion can then be confirmed by a scientist, which subsequently enables the AI to better identify species throughout the world. The data collected also allow scientists to track changes in terms of species population and abundance, receive almost real-time pest invasion information, and apprehend how plants and animals adapt to changes such as climate change or desertification.[97]

The civil society organization (CSO) Wild Me developed a new collaborative platform called Wildbook to help scientists benefit from citizen science, computer vision, and AI, and generate new insights about wildlife species. This initiative is a collaboration between Wild Me and the University of Illinois-Chicago, Rensselaer Polytechnic Institute, Princeton University, but also Microsoft, Pineapple Fund, and Amazon Web Services. Wildbook offers technological solutions to collect and store large volumes of data originating in the field. It allows scientists to monitor individual animals and their behavior in the wild, collect and analyze biological samples, engage citizen conservationists, build a research network, and create new animal biometrics.[98] Thanks to AI, images showing animals with distinctive features such as stripes, spots, wrinkles, or notched markings can easily be identified and large-scale databases about wildlife population movements at specific times of the year can be generated. In addition, AI allows these data from the field to be combined with geographic, environmental, behavioral, and climate data, affording better comprehension of what animals do, when, where, and why.

Wildbook allows individual animals in photos collected by citizen conservationists to be recognized through two stages. First, the detection process, based on deep convolutional neural networks (DCNNs), detects animals and some of their specific features, creates bounding boxes surrounding each of them, and annotates the bounding boxes with the description of these features. Second, the identification process assigns a name to the annotations generated by the detection phase. The specific features of each animal and their location on its body produce a specific score for each individual animal.[99] Wildbook supports a wide range of conservation projects and institutions, including WWF Norppa Galleria for Saimaa ringed seals, MantaMatcher for manta rays, Giraffespotter for collaborative giraffe research, Princeton University for zebras and giraffes, and SPLASH Catalog for humpback whales (*Megaptera novaeangliae*), to name but a few.[100]

Developed by the same CSO Wild Me, the Wildbook for Whale Sharks is a project that aims at identifying whale shark (*Rhincodon typus*) encounters and individually catalogued whale sharks. Through citizen science, photos of whale sharks are taken, with observation data such as location, time, sex, and scars. These data are verified by a scientist, and then uploaded into the database. Two spot pattern-matching algorithms, functioning similarly to facial recognition software, are used to analyze the skin patterning behind the gills of each animal, including any existing scars. This analysis provides a ranked selection of possible matches with existing whale sharks in the database, and if there is no match, a new profile is created.[101]

Another aspect of knowledge production is related to reporting on and monitoring compliance. Indeed, assessment is often challenging in global environmental governance due to its complexity and large number of conventions, treaties, and the like. Many governmental agencies and IOs have their own reporting formats, which leads to a juxtaposition of different reporting formats and data sets for the same agreement or the same objective. In the field of biodiversity conservation, more than 155 multilateral and bilateral agreements cohabit, each one with its own reporting and monitoring process. This makes monitoring highly complicated. AI technology can prove useful to solve this challenge thanks to its capacity to analyze large volumes of structured but also unstructured data. It can help CSOs provide a global overview of all activities and reporting for a global issue such as biodiversity. It can help better assess where resources are spent for a global objective, and based on prediction models, enhance how they will be allocated in the future. This lends more efficiency and effectiveness to donor funding, making recipients of these funds more accountable as well.[102]

In a digitalized world, criminals use social media platforms to illegally trade tiger products, such as their teeth, claws, parts of their skin, and other parts of this animal that are used in traditional medicine. The World Wide Fund for Nature (WWF) has collaborated with the company Tiger Beer, 13 Asian countries, and six artists[103] to counterattack, using the same channel (social media), and they have launched a communication campaign entitled #3890Tigers campaign – 3,890 refers to the estimated number of tigers left in the world today. The objective of this initiative is to raise awareness of the declining population of tigers in the wild. AI is used in this case to transform pictures and selfies that people have taken into unique artwork. Each portrait is intertwined with a tiger to showcase how people can live harmoniously with tigers. This artwork is then shared on social media, which enables WWF to give this cause visibility. The communication campaign pledges to end the illegal tiger trade and aims at doubling the tiger population by 2022.[104]

5. Concluding remarks

With about two billion 1-megapixel photographs produced every day,[105] collecting data from satellites is no longer an issue. To the contrary, the volumes of data collected require the use of specific technologies to analyze these large data sets and make sense of them, both for the global good and to protect the planet. According to Stuart Russell, UC-Berkeley, lead of the AI for Good breakthrough team on AI and satellite imagery, "We have recorded the whole world for a long time," which has led to having access to this unprecedented data set. However, due to the complexity of the world, humans have a hard time making sense out of it. However, "with AI, perhaps we can."[106]

AI is increasingly taking a leading role in making sense of the big data recorded by satellites, drones, smartphones, and sensors throughout the planet. Indeed, this technology can be applied to help solve a large number of contemporary challenges. As discussed in this chapter, AI supports civil society in its roles of service provider and knowledge broker. It allows the development of innovative solutions to combat poaching and better protect endangered species and ecosystems, among others. It also allows the production of new knowledge about life on land and under the seas, and a more precise understanding of our planet and its environmental complexities.

However, civil society is not the only actor of global environmental governance to benefit from this technology: local public authorities, businesses, and international organizations also embrace AI to improve their services and develop new knowledge.

AI can empower SDG Goal 11 entitled "Make cities inclusive, safe, resilient and sustainable"; autonomous public transportation, efficient resource allocation, urban planning based on predictive models combining satellite imagery, economic activities, transportation flows, private consumption, entertainment and recreation activities, and crime rates are some examples of the application of AI to develop smart and sustainable urban areas.[107] Furthermore, by integrating data from production and consumption on a large scale and with real-time data, AI can improve the efficiency of food production, distribution, and consumption, reduce waste, and reach SDG Goal 12, "Ensure sustainable consumption and production patterns."[108]

Renewable sources of energy are intermittent. Thanks to real-time data collection and analysis, AI can develop accurate consumption prediction models, and thus increase the use of green energy and its integration into the traditional electric grid, supporting SDG Goal 7 entitled "Ensure access to affordable, reliable, sustainable and modern energy for all."[109] Furthermore, big data and the Internet of Things (IoT) generate large volumes of data. Analyzed by AI, the consumption patterns allow accurate sanitation predictions to be made based on real-life and real-time data and not models from past years. Thanks to these predictions, clean water distribution and sanitation can be planned more accurately, in support of SDG Goal 6 entitled "Ensure access to water and sanitation for all."[110]

Artificial intelligence can help improve the integration of supplies of renewable energy into the existing electricity power grid. Indeed, solar, wind, hydroelectric, biomass, and geothermal power are some of the most well-known sources of renewable energy. However, they provide energy irregularly contrary to gas or nuclear power plants for instance, which can be managed according to the electricity consumption demand. This irregularity is one of the main challenges associated with renewable energies, and AI can help overcome this issue by providing almost real-time control over the supply and demand of electricity. Indeed, by 2020, the European Union aims to implement more than 200 million smart electricity meters for private consumption. These meters are connected to the electric grid and provide real-time data about individual consumers. These large volumes of data are collected and then analyzed by AI to develop electricity demand prediction models based on consumer constraints and preferences. Based on this analysis, AI can adapt the supply of electricity almost in real time. Indeed, smart meters can reduce the electricity provided by temporarily dimming lights or switching off electric heaters. The smart meter functionalities, combined with AI, allow the integration of a larger share of intermittent renewable sources of energy into the electric grid, leading to more sustainable energy production and consumption.[111]

AI can also support communities and cities in increasing their efficiency when providing public services. For instance, the city of Melbourne in Australia is developing an AI-based platform that in real time can adapt the delivery of clean water to the demands for clean water that are coming, also in real time, from the inhabitants of the city.[112] This leads to reducing by about 20% the amount of energy used to treat water. Python, the name of the platform, combines historical and real-time consumption data to ascertain the most efficient use of water pumps without any human intervention.[113] Started as a pilot in one water treatment plant, it is rapidly expanding to others in the city. To ensure that the delivery of water would continue even if there were an incident, the AI can only optimize the process. In addition, Melbourne Water has implemented cybersecurity provisions, one of which is that the AI platform is not connected to a wider network.[114]

AI has multiple applications to help reach SDG Goal 16 entitled "Promote just, peaceful and inclusive societies." It can help organizations develop new services and provide broader access to e-government services and information thanks to a smart and automated voice assistant.[115] It can also help improve trust between various stakeholders. The International Telecommunication Union (ITU), along with a network of other international organizations, has developed a platform called TrustFactory.ai.[116] This platform is an incubator to support projects on three aspects related to trust and AI: trust in AI technologies, trust in AI developers, and trust among users of AI technologies, which includes a large number of stakeholders such as developers, policy and decision makers, local communities, NGOs, businesses, and individuals, to cite only a few.[117]

Although AI presents some challenges in terms of governance and power – the fear of humans of being controlled by AI one day – this new technology already allows us to better understand our planet, from its urban communities to its most remote and untouched areas, from high mountain peaks to deep oceans, and to understand how our planet adapts to the rapid changes it must face stemming from human activities, climate change, and pollution, to cite only a few.

This chapter aimed at providing an overview of the current uses of AI in relation to nature conservation and sustainable development. It is not exhaustive by far. Due to the fast-changing environment of this disruptive technology, AI will most certainly become a major instrument in the future of environmental governance, from policy making to field and operational projects. The next challenge for the environmental community and society at large is to develop AI skills in order to benefit the most from this technology and help protect the planet.

Notes

1 Cellan-Jones, R., 2014. *BBC*. www.bbc.com/news/technology-30290540
2 Ibid.
3 Ibid.
4 Holley, P., 2018. Elon Musks nightmarish warning AI could become an immortal dictator from which we would never escape. *The Washington Post*. www.washingtonpost.com/news/innovations/wp/2018/04/06/elon-musks-nightmarish-warning-ai-could-become-an-immortal-dictator-from-which-we-would-never-escape/
5 Vincent, J., 2017. Putin says the nation that leads in ai 'Will be the ruler of the world. *Verge*. www.theverge.com/2017/9/4/16251226/russia-ai-putin-rule-the-world
6 Clark, C., 2017. Our artificial intelligence 'Sputnik Moment' is now: Eric Schmidt & Bob Work. *Breaking Defense*. https://breakingdefense.com/2017/2011/our-artificial-intelligence-sputnik-moment-is-now-eric-schmidt-bob-work/
7 Koch, M., 2018. Artificial intelligence is becoming natural. *Cell*, 173(3), pp. 531–533, 531. https://doi.org/10.1016/j.cell.2018.04.007
8 Horowitz, M.C., 2018. Artificial intelligence, international competition, and the balance of power. *Texas National Security Review*, 1(3), p. 39. https://tnsr.org/2018/05/artificial-intelligence-international-competition-and-the-balance-of-power/
9 CNBC, 2018. *Humanoid Robot Sophia – almost human or PR Stunt*. www.youtube.com/watch?v=7fnCQC7bLs0
10 Good Morning Britain, 2017. *Humanoid Robot Sophia tells jokes on GMB*. www.youtube.com/watch?v=kWlL4KjIP4M
11 Tegmark, M., 2017. *Life 3.0: Being human in the age of artificial intelligence*, 1st ed. New York, NY: Knopf.
12 Russel, S., 2017. Artificial intelligence: The future is superintelligent. *Nature*, 548, pp. 520–521.
13 Bostrom, N., 2014. Clever cogs: The potential impacts of intelligent machines on human life. *The Economist*. www.economist.com/books-and-arts/2014/08/09/clever-cogs
14 Silver, D., Hassabis, D., 2016. Google research blog: AlphaGo mastering ancient game of go. *Google Research Blog*. http://googleresearch.blogspot.co.uk/2016/01/alphago-mastering-ancient-game-of-go.html.
15 Gurkaynak, G., Yilmaz, I., Haksever, G., 2016. Stifling artificial intelligence: Human perils. *Computer Law & Security Review*, 32, p. 749. http://dx.doi.org/10.1016/j.clsr.2016.05.003
16 Cadwalladr, C., 2014. Are the robots about to rise? Google's new director of engineering thinks so. . . *The Guardian*. www.theguardian.com/technology/2014/feb/22/robots-google-ray-kurzweil-terminator-singularity-artificial-intelligence

17 Faremo, G., 2018. Artificial intelligence can help achieve The SDGs. *Speech at the Second Annual Digital Workforce Summit*, New York, NY, 7 June 2018. www.unops. org/news-and-stories/speeches/the-second-annual-digital-workforce-summit

18 Garnham, A., 2018. *Artificial intelligence: An introduction*. London, UK: Routledge & Kegan Paul, p. 3.

19 McClelland, C., 2017. The difference between artificial intelligence, machine learning, and deep learning. *Medium*. https://medium.com/iotforall/the-difference-between-artificial-intelligence-machine-learning-and-deep-learning-3aa67bff5991

20 Garnham, A., 2018. *Op Cit.*, p. 3.

21 Bhatia, R., 2017. Understanding the difference between Symbolic AI & Non Symbolic AI. *Analytics India*. https://analyticsindiamag.com/understanding-difference-symbolic-ai-non-symbolic-ai/

22 Campbell, M.A., Hoane, J.J., Feng-hsiung, H., Deep Blue. *Artificial Intelligence*, 134(1–2), pp. 57–83. https://doi.org/10.1016/S0004-3702(01)00129-1

23 Smolensky, P., 1987. Connectionist AI, symbolic AI, and the brain. *Artificial Intelligence Review*, 1(2), p. 95. https://doi.org/10.1007/BF00130011

24 Bhatia, R., 2017. *Op Cit.*

25 Urban, T., 2015. The AI revolution: The road to superintelligence. *Wait but Why Blog*. http://waitbutwhy.com/2015/01/artificial-intelligence-revolution-1.html

26 Gurkaynak, G., Yilmaz, I., Haksever, G., 2016. *Op Cit.*, p. 751.

27 Ibid.

28 Donald Knuth quoted by Nilsson, N.J., 2010. *The quest for artificial intelligence*. New York, NY: Cambridge University Press.

29 Goertzel, B., Pitt, J., 2012. Nine ways to bias open-source AGI toward friendliness. *Journal of Evolution and Technology*, 22(1), pp. 116–131.

30 Bostrom, N., 2006. How long before superintelligence? *Linguistic and Philosophical Investigations*, 5(1), pp. 11–30.

31 Yudkowsky, E., Salamon, A., Shulman, C., Kaas, S., McCabe, T., 2010. Reducing long-term catastrophic risks from artificial intelligence. *Intelligence.Org*. https://intelligence.org/files/ReducingRisks.pdf

32 Gurkaynak, G., Yilmaz, I., Haksever, G., 2016. *Op Cit.*, p. 752.

33 Domingos, P., 2015. *The master algorithm: How the quest for the ultimate learning machine will remake our world*. New York, NY: Basic Books, p. 1.

34 Ibid, p. 2.

35 Ibid, p. 3.

36 Ibid, p. 5.

37 Reynolds, M., 2017. How to explain AI to your family over the Christmas turkey. *Wired*.www.wired.co.uk/article/explain-artificial-intelligence-to-your-family-machine-learning

38 McClelland, C., 2017. The difference between artificial intelligence, machine learning, and deep learning. *Medium*. https:// medium.com/iotforall/the-difference-between-artificial-intelligence-machine-learning-and-deep-learning-3aa67bff5991

39 Accenture, 2018. *AI explained: A guide for executives*, p. 6. https://view.pagetiger.com/AI-Explained-A-Guide-for-Executives/2018/page6.htm

40 Rintanen, J., 2015. *A brief overview of AI planning*. https://users.aalto.fi/~rintanj1/planning.html

41 Techopedia,2018.*What is machine perception?* www.techopedia.com/definition/32792/machine-perception

42 Stanek, M., 2017. Moravec's paradox. *Medium*. https://medium.com/@froger_mcs/moravecs-paradox-c79bf638103f

43 For more information on affective computing. www.media.mit.edu/groups/affective-computing/overview/

44 The terms "AI" and "artificial intelligence" will be used as an umbrella term to describe the different technologies it encompasses.

45 Bondi, E., Dey, D., Kapoor, A., Piavis, J., Shah, S., Fang, F., et al., 2018. AirSim-W: A simulation environment for wildlife conservation with UAVs. *Compass '18: ACM*

SIGCAS conference on computing and sustainable societies (COMPASS), 20–22 June 2018, ACM, Menlo Park and San Jose, CA, New York, NY, p. 12. https//doi.org/10.1145/3209811.3209880

46 Xprize. *Op Cit.*

47 ITU, 2018. *AI breakthrough tracks.* www.itu.int/en/ITU-T/AI/2018/Pages/break-through-tracks.aspx#04

48 Ibid.

49 Ibid.

50 Goal 15. *Protect, restore and promote sustainable use of terrestrial ecosystems, sustainably manage forests, combat desertification, and halt and reverse land degradation and halt biodiversity loss.* See more https://unstats.un.org/sdgs/indicators/indicators-list/

51 Microsoft, 2018. *Advanced mapping for precision conservation.* www.microsoft.com/en-us/aiforearth/land-cover-mapping.aspx

52 Snow, J., 2016. Rangers use artificial intelligence to fight poachers. *National Geographic.* https://news.nationalgeographic.com/2016/06/paws-artificial-intelligence-fights-poaching-ranger-patrols-wildlife-conservation/

53 McLean, A., 2017. Neurala partners with Lindbergh Foundation to use drones to combat poaching. *ZDNet.* www.zdnet.com/article/neurala-partners-with-lindbergh-foundation-to-use-drones-to-combat-poaching/

54 Harvey, L., 2017. Deep learning aids wildlife conservation by air, land, and sea. *MathWorks Blogs.* https://blogs.mathworks.com/headlines/2017/05/30/deep-learning-aids-wildlife-conservation-by-air-land-and-sea/

55 McLean, A., 2017. *Op Cit.*

56 Smith, T., 2017. Protecting wildlife with AI. *Twenty-One Twenty-Eight.* www.twentytwotwentyeight.com/single-post/2017/07/15/Protecting-Wildlife-with-AI

57 Goodchild van Hilten, L., 2016. AI springs into action in surprising places. *Elsevier Connect.* www.elsevier.com/connect/using-ai-and-game-theory-to-outwit-poachers

58 Smith, T., 2017. *Op Cit.*

59 Snow, J., 2016. *Op Cit.*

60 Fang, F., Nguyen, T., Ford, B., Sintov, N., Tambe, M., 2015. Introduction to green security games (Extended Abstract). *USC.* http://teamcore.usc.edu/papers/2015/IJCAI2015_GSG.pdf

61 Ibid.

62 Snow, J., 2016. *Op Cit.*

63 Ibid.

64 Yong, X. How artificial intelligence is revolutionising conservation. *This Blog.* Deakin University. http://this.deakin.edu.au/innovation/how-artificial-intelligence-is-revolutionising-conservation

65 BeeScanning App. https://beescanning.com/eng/

66 Harvey, L., 2017. *Op Cit.*

67 Digital journal, 2017. *The beescanning app is saving bees worldwide through deep learning technology.* www.digitaljournal.com/pr/3347240#ixzz5NPMGSiLE

68 Ibid.

69 BeeScanning app. https://beescanning.com/eng/faq-eng/

70 ITU, 2018. *AI breakthrough tracks.* www.itu.int/en/ITU-T/AI/2018/Pages/break through-tracks.aspx#04

71 Xprize. *Op Cit.*

72 Ibid.

73 United Nations Sustainable Development Goals (SDGs). www.un.org/development/desa/disabilities/envision2030.html

74 ITU. *AI breakthrough tracks.* www.itu.int/en/ITU-T/AI/2018/Pages/breakthrough-tracks.aspx#04

75 Microsoft, 2018. *Data-driven farming to sustainably feed the world.* www.microsoft.com/en-us/aiforearth/farmbeats.aspx
76 Ibid.
77 Xprize, 2018. *AI solving sustainable development goals.* https://ai.xprize.org/AI-For-Good/sustainable-development-goals
78 Project Premonition. www.microsoft.com/en-us/aiforearth/project-premonition.aspx
79 Microsoft. www.microsoft.com/en-us/aiforearth/projects.aspx
80 Xprize. *AI solving sustainable development goals.* https://ai.xprize.org/AI-For-Good/sustainable-development-goals
81 Catlin Seaview Survey. http://catlinseaviewsurvey.com
82 Catlin Seaview Survey. http://catlinseaviewsurvey.com/partners
83 Catlin Seaview Survey. http://catlinseaviewsurvey.com
84 Tollefson, J., 2016. Computers on the reef: Software tools that digitize and annotate underwater images are transforming marine ecology. *Nature.* www.nature.com/news/computers-on-the-reef-1.20497
85 Ibid.
86 CoralNet. UC San Diego. https://coralnet.ucsd.edu/about/
87 Tollefson, J., 2016. *Op Cit.*
88 The Nature Conservancy, 2016. *How tech is saving our oceans.* www.nature.org/ourinitiatives/urgentissues/oceans/providing-food-sustainably/fishface-using-technology-to-change-the-way-fisheries-are-managed.pdf
89 Harvey, L., 2017. Deep learning aids wildlife conservation by air, land, and sea. *MathWorks Blogs.* https://blogs.mathworks.com/headlines/2017/05/30/deep-learning-aids-wildlife-conservation-by-air-land-and-sea/
90 Briony, H., 2018. The latest weapons in the fight against ocean plastic? Drones and an algorithm. *WEF.* www.weforum.org/agenda/2018/06/this-ai-is-learning-to-recognize-ocean-plastic-using-drone-photos
91 The Plastic Tide. www.theplastictide.com
92 Briony, H., 2018. How AI-powered drones are helping fight ocean plastic. *ITU.* https://news.itu.int/ai-drones-ocean-plastic/
93 Xprize. *Op Cit.*
94 Ibid.
95 Wildbook: software to combat extinction. www.wildbook.org/doku.php#publications
96 iNaturalist. www.inaturalist.org/pages/what+is+it
97 Microsoft for Earth. iNaturalist. www.microsoft.com/en-us/aiforearth/inaturalist.aspx
98 Wildbook: software to combat extinction. www.wildbook.org/doku.php#publications
99 Ibid.
100 Ibid.
101 Whale shark. www.whaleshark.org
102 Balakrishna, 2018. Big data, block chains, artificial intelligence: The new future for conservation. *Fledgein.* http://fledgein.org/bigdata-ai-conservation/
103 Smith, T., 2017. Protecting wildlife with AI. *Twenty-One Twenty-Eight.* www.twentytwotwentyeight.com/single-post/2017/07/15/Protecting-Wildlife-with-AI
104 WWF. #3890tigers project. http://wwf.panda.org/get_involved/partner_with_wwf/business_partnerships/tiger_beer_partnership/
105 ITU, 2018. *AI and satellite imagery: Proposed 'global service platform' to scale AI for good projects.* https://news.itu.int/ai-and-satellite-imagery-proposed-global-service-platform-to-scale-ai-for-good-projects/
106 Ibid.
107 Xprize. *Op Cit.*
108 Ibid.
109 Ibid.

110 Ibid.
111 Robu, V., Flynn, D., 2017. Artificial intelligence: Outsmart supply dips in renewable energy. *Nature*, 544, p. 161. www.nature.com/articles/544161b
112 ITU, 2018. *How Melbourne, Australia uses AI to cut water treatment costs.* https://news.itu.int/melbourne-cut-down-water-costs-using-ai/
113 Wells, H., 2018. Melbourne cuts water treatment costs using AI. *Cities Today.* https://cities-today.com/melbourne-cuts-water-treatment-costs-using-ai/
114 Ibid.
115 Xprize. *AI solving sustainable development goals.* https://ai.xprize.org/AI-For-Good/sustainable-development-goals
116 TrustFactory.ai. https://trustfactory.ai/about-1/
117 ITU Breakthrough Tracks. Track 4: Trust. www.itu.int/en/ITU-T/AI/2018/Pages/breakthrough-tracks.aspx#04

Concluding remarks

This book was motivated by a deep concern about the urgent necessity to protect nature and to find a global response to this need. Today's world is organized around information and data, with ongoing technological, industrial, organizational, and commercial innovations; an increased level of complexity due to the multiplicity of forms of authority and the interdependence of global issues; and an unforeseen number of changes in the environment – climate and biodiversity loss, to cite only two. These new technological and environmental conditions raise many concerns, starting with their global scope and their speed. In this sense, technology and the environment could be seen as similar: they change fast, and their impact is at all levels – local, national, regional, and international – and on all sectors of the economy and on all social-economical groups of society. However, the impact is not homogenous for all and varies greatly from sector to sector, from one country, region, or community to another, and from one social-economical group of society to another. Indeed, the capacity to adapt to these changes depends on the capacity to adapt to environmental changes such as rising sea levels or higher temperatures, and to technological changes such as the emergence of new jobs and the disappearance of others.

This complexity calls for new collective governance procedures that allow all stakeholders not only to be part of the rule-making processes, but also to gain additional ownership of the implementation of these decisions. The responsibility of governing the environment and digital technologies is at all levels, not only among actors on the international stage but also among citizens who use these digital tools, and whose daily choices and behaviors have a direct impact on the environment. In this sense, new collective governance procedures also imply that all actors, from the citizen to the state representative, perceive their responsibility to tackle these issues. Since the generalization of ICTs did not lead to the emergence of a global citizen identity, maybe the challenges to managing common global public goods, such as technologies and the environment, could lead to a global understanding of a common fate requiring people to face these global challenges, whose impacts on society and institutions are yet to be understood fully.

International decision-making mechanisms face the challenge not only of legitimacy, but also of efficiency and effectiveness. In that respect, digital technologies can prove to be a double opportunity. On one hand, secretariats and organizations can become more efficient by using ICTs to improve their internal processes and reduce their communication costs. On the other hand, ICTs can also replace some of their tasks, in particular the most routine and administrative tasks, which will require, for some of them, showing flexibility and adaptation, but also having the autonomy to redefine their role and added value on the international stage. This is an opportunity since it will allow more resources – human and financial – to be dedicated to field projects and research. In that respect, ICTs could help bridge the gap between the mandate given by state parties and the actual concrete resources. In that sense, ICTs could help organizations to become more effective and visible.

In the early 2000s, when social media platforms emerged, optimists would praise how digital technology will change the organizations through enhanced communication tools, and empower individuals and civil society, raising awareness and spreading democracy throughout the world. At the same time, pessimists already stressed the repercussions of technological advancements: tottering digital security, and increasing inequality – especially due to the unequal internet penetration rates. Today, digital technologies and the challenges they represent have emerged in the public debate, and the pessimist view seems to have won the public debate. While more and more citizens use social media platforms, the impact of ICTs on political processes, in particular democratic political processes, raises a lot of concerns.

While technological determinism recognized the absolute power of ICTs to transform societies into democracies, the Arab Spring and Winter, among other examples, showed that this form of determinism is inaccurate[1]. Digital technologies have an impact on societies, but not in such a deterministic manner. And their impact can be positive, improving the communication skills of civil society for instance, while it can also be negative, by propagating fake news and influencing the votes of millions of citizens through the manipulation of facts. If this book focuses on the positive side of digital technologies, it also acknowledges the danger they represent when not well governed. Hence, these technologies require, as does the environment, efficient and effective governance mechanisms for us to benefit the most from them. States have perceived the opportunities, but also the menaces, that this new digital online space – also called cyberspace – represents, and have developed new cyber-capacities to compete and collaborate. Ongoing cyber-attacks, where states fight against each other to advance their interests, and criminal groups that pursue criminal activities and make immense profits are additional illustrations of the impact of ICTs on the international stage and on the individual lives of citizens, which calls for better governance of ICTs.

In addition, ICTs are not neutral. They are developed in a specific cultural context, and reproduce the values, customs, and habits of the country they come from.[2] The idea of using ICTs to empower civil society probably stems from an ethnocentric view, where civil society's participation in local to international decision-making processes is perceived as positive. However, this view is not universal and raises multiple questions concerning the representability of civil society organizations. Indeed, CSOs pursue not only substantive goals, such as the protection of the environment or the representation of the interests of local and indigenous communities, but also organizational goals, including their growth, influence, and autonomy.[3] Hence, their use of ICTs reflects these two categories of goals, and their activities on the international stage as well. Necessarily, and similarly to other actors, they need to raise additional funds, and attract the attention of citizens at the local level and governmental actors on the international stage. Their legitimacy stems from multiple sources, starting with their scientific expertise and knowledge of the field. The environmental CSOs mentioned in this book have a long history of conservation projects, advocacy campaigns, and monitoring undertakings. Although some critics may raise some concerns about the category of goals they pursue – substantial or organizational – their accomplishments are well-established and have led to the emergence of the legal framework to protect the environment that we know today, and to the better protection of species and ecosystems. Moreover, their legitimacy comes from the large community of citizens who actively support their work. Although CSOs are not elected, it is undeniable that they have represented the interests of parts of the populations – and of the world – that were once not as well represented on the international stage as they are today. In 2016 IUCN Members decided to create a new category of Members dedicated to indigenous people and communities, which is one illustration among others of the crucial influence of civil society on international decision-making processes. Hence, ICTs help CSOs gain additional legitimacy by allowing a closer bond between CSOs and the ones they claim to represent, and by improving their efficiency and effectiveness both in terms of advocacy and in terms of conservation field projects.

This book focused on the use of current and emerging ICTs by environmental civil society organizations, and how they empower these organizations to reinforce their competences to participate in global environmental governance. Although the international stage continues to be based on the state system, civil society organizations make extensive use of ICTs, and in an information age, their capacity to use these current and emerging technologies leads them to gain more visibility and credibility.

The participation of civil society and other stakeholders as well as the role of technologies are considered to be already in the premises of the emergence of the concept of sustainable development. Civil society rose as a global movement of actors on the international stage in parallel to the generalization of ICTs and globalization. These three aspects of our contemporary world are intrinsically connected. The emergence of new forms of authority accentuates the complexity of global environmental governance, while enriching the decision-making processes with a variety of points of view and interests, scientific expertise, and local knowledge. In this informational society, with information and data at its core, the Net generation shows unprecedented levels of ICT mastery since its members were born in an internet age. In this context, where transparency, cooperation, and participation are the new normal for this generation, civil society adapts by developing new forms of governance without institutionalization, and new forms of technology-intensive activism. These new forms of action on the international stage, combined with an increased access to information, better education, and an increased awareness of global issues such as the environment, represent a new opportunity for the protection of the environment.

Among the numerous tools used by civil society to reach out to its audiences, websites are the oldest ones. The analysis of 15 CSOs accredited to UNEA indicates that they first and foremost use their websites to provide information to a wide audience. However, few have developed consultation and mobilization features on their websites. This analysis also shows that web traffic is a strong indicator of the number of participation features on their websites. CSOs with high web traffic have developed substantially more information, consultation, and action features on their websites, which also show higher levels of usability and maturity. Conversely, organizations with less web traffic have developed less content and implemented fewer technology features. This leads one to conclude that web traffic is an indicator of participation, either as an incentive to develop participation features or as a consequence of the presence of such features on the website. It seems, in any case, logical to pursue the idea that a website, with well-developed content, along with numerous consultation and action initiatives, accessibility and ease-of-use, and high sophistication, can only increase opportunities for transparency and interaction. However, among the endless number of websites available, only a few get referenced on the first page by search engines. This means that for smaller organizations with a limited online advertising budget, a website might in fact not result in more visibility. When searching for key words such as "environment," "climate change," or "biodiversity," the main international governmental and non-governmental organizations come first and get most of the attention. This leads one to conclude that although websites are in principle extraordinary tools to promote content and the work of an organization, they can also reinforce the inequalities between small organizations and the big ones.

The second type of digital technology most used by civil society is social media. At a time when information is increasingly consulted on social media first, and when individuals spend more and more time connected and in front of a screen, these platforms are crucial for civil society to increase its visibility and reach out to its audience. The analysis of the social media presence of some CSOs accredited to UNEA confirms the dominant narrative that hints that social media instruments have become an integral part of the advocacy strategies of CSOs. A large majority of organizations are present on Facebook and Twitter. The digital divide remains an issue with the number of NGOs from Europe and North America surpassing African ones by far. Moreover, and similarly to websites, social media tends to increase the inequality between small and large organizations, with the top 15 organizations with the most likes (on Facebook) and followers (on Twitter) concentrating most of the visibility. This implies that a fairly small number of organizations, mainly from Europe and North America, concentrate most of the attention on these two platforms, and therefore inhibit organizations from other continents from raising awareness about other issues and proposing other perspectives about nature conservation. It shows that social media platforms have not only allowed a large number of organizations to reach out to global audiences, but also contributed to inhibiting smaller organizations from raising awareness about other issues and to reducing the plurality of sources of information for citizens.

Furthermore, the choice of advocacy tactics, tone of voice, regularity of publishing, and story line depends on how organizations are positioned in global environmental governance. Their communications were coherent throughout the month chosen for the analysis, and clearly illustrate the variety of advocacy strategies that are possible thanks to social media platforms. Finally, this analysis also partly confirms previous research, in the sense that the sample of CSOs examined mainly used social media to inform their audience. As Lovejoy explains, "Although nonprofit organizations have become more interactive in their use of Twitter as opposed to their websites alone, we found Twitter is still used by many nonprofit organizations as an extension of information-heavy websites. These organizations are missing the bigger picture of its uses as a community-building and mobilizational tool."[4] However, the call to action comes second in terms of social media tactics, and many organizations use these platforms to mobilize their audience and involve them to support their work, change a behavior, take part in local projects and protests, or increase the awareness of an issue. A similar analysis of other platforms used in Asia would reveal highly relevant information.

With the emergence of civil society as actors in global environmental governance, ICTs are increasingly used to foster decision-making processes. This was one of the main promises of the internet: it was supposed to allow new stakeholders to take part in governance processes at all levels since the internet allows the communication of the many to the many. The multi-stakeholder decision-making process developed by the International Union for Conservation of Nature (IUCN) gives the same voting rights to state and civil society Members, and enables Members to make proposals in the field of nature conservation at large. Its progressive digitalization offers an opportunity to analyze if ICTs can improve the participation of stakeholders, and in particular civil society, in a decision-making process. The online discussion platform and the online voting procedures allowed Members to reach a consensus online on more than 90% of the topics debated. Furthermore, the analysis shows that this digital process allowed state and non-state Members to participate at similar levels: in other words, it did not favor the most resourceful actor and succeeded in providing equal opportunities to both categories of Members. If this case of digital participation is conclusive, it is also due to two factors associated with technology adoption. Indeed, participating in the electronic discussion system required Members to adopt a new technology. This is not as straightforward as it may seem. Indeed, a barrier often prevents users from adopting a new technology, whether this is due to its complexity or to the user's limited understanding, competence, or time. This barrier can be either psychological or technological. For both cases, the electronic discussion system was successful in developing a technology that was accessible to all constituents and could be easily used. Furthermore, to adopt a new technology, overcome the barriers mentioned previously, and change the habit of discussing and debating motions from on site to online require clear and direct motivation. This means that in the case of the electronic discussion system, Members clearly saw the direct benefits of adopting this new technology. Lastly, the digitalization process of this governance mechanism allows the organization to reduce significantly its environmental footprint, since it allows participants to make propositions and take decisions remotely.

Further research to assess the impact of ICTs on other global decision-making mechanisms is required and would help to gain additional understanding of their potential impact. In addition, an analysis of all forms of civil society participation and consensus-building processes in global environmental governance could help bring forward general trends and show the role of new ICTs in this evolution. In particular, additional research should be conducted to examine the influence of civil society organizations on treaties and global agreements to protect nature. If some global environmental governance mechanisms are more legitimate than others, the next stage would be to define if the global civil society has gained more influence on the final version of treaties because of new ICTs. Moreover, levels of participation in the discussion process were low and average in the voting process. Further research should enable one to understand how to increase these levels of participation.

In terms of emerging technologies, blockchain is one of the most prominent, with numerous articles and videos online mentioning its benefits for people and organizations. This global digital ledger offers innovative solutions in terms of trust, fundraising, transparency, incentives, and distributed governance. Blockchain technology provides a new set of skills to CSOs. It has the potential to positively affect any type of decision-making process involving multiple parties globally. Good governance is key to managing natural resources sustainably. Values of trust, transparency, inclusive participation, and effective implementation are the building blocks of future global governance systems that will ensure a healthy and prosperous future for all. Furthermore, blockchain technology can increase the visibility of sustainable and unsustainable production practices globally. This would help consumers to make a choice when buying products and services. The network itself, thanks to its structure, ensures trust among all agents and allows information such as land property rights to be recorded safely. Local communities with rights to natural resources could receive direct payments in bitcoins as a reward and incentive to protect their nearby ecosystems and species. However, this technology also raises numerous concerns, including data security and privacy, the right to be forgotten, high levels of computing capacity and therefore a high level of electricity consumption, and finally access to this technology, which remains unequally distributed in the world.

As mentioned previously, information and data are at the center of the knowledge economy and the informational society. Individuals, organizations, and governments produce and consume large amounts of data. Big data allow CSOs to develop a wide range of innovations in fields relating to global environmental governance. The analysis focused on the knowledge production and distribution capacity of CSOs. As shown, civil society is increasingly becoming tech-savvy and gradually embracing these emerging technologies to increase its positive impact on the protection of the environment, the combat against climate change, and the loss of biodiversity, to name only few. However, big data technologies are not without critics, in particular with regard to questions of privacy and confidentiality, compatibility of data sets, and biases. Indeed, these large data sets are not always based on a common standard, which reduces their impact and opportunities for innovation and nature conservation. Moreover, data collected by citizens can be inherently and unintentionally biased, which leads researchers to focus more attention on data integrity. Due to their intrinsic features, big data require new non-human instruments and techniques in order to be handled properly.

Big data technologies are linked to another innovation, which is increasingly used by civil society in the environmental field: artificial intelligence (AI). It is increasingly taking a leading role in making sense of the big data recorded by satellites, drones, smartphones, and sensors throughout the planet. Indeed, this technology can be applied to help solve a large number of contemporary challenges, and enhance organizations' knowledge production and field implementation capacity. AI can help in predicting the effects of climate change and the path of storms, monitoring the spread of diseases, understanding human decision-making processes in terms of energy consumption and protection of the environment, or identifying areas most in need of attention. These are some of the numerous AI applications that CSOs have employed in the last several years to protect the planet. Although AI presents numerous challenges in terms of ethics, governance, and power, this new technology already allows us to better understand our planet, from its urban communities to its most remote and untouched areas, from high mountain peaks to deep oceans, and to understand how our planet adapts to the rapid changes it must face stemming from human activities, climate change, and pollution, to cite only a few. AI will most certainly become a major instrument in the future of environmental governance, from policy making to field and operational projects.

In this fast-changing context, where innovations keep emerging and triggering numerous others, this book aims at providing an overview of the current uses of current and emerging ICTs by CSOs in global environmental governance. It does not aim to provide an exhaustive list. However, it wishes to show that digital technologies can be beneficial to environmental CSOs but at different degrees. In a world with information at its core; citizens who spend an increasing amount of time consuming and producing information online; and citizens who are more aware of environmental challenges; current and emerging ICTs allowing organizations to increase their visibility, to advocate, to engage with local communities, and to conduct scientific experiments and field projects, civil society has the opportunity not only to pursue its key roles in global environmental governance, but to enhance some of its key competences, including advocacy, making proposals, fundraising, promoting sustainable behaviors, monitoring, knowledge production and distribution, and field project implementation. Hence, digital technologies strengthen the participation of CSOs in global environmental governance.

This book aims to contribute to building additional knowledge on the prevalence, limits, and opportunities of new ICTs. It wishes to show how current and emerging ICTs can be deployed to support global environmental governance, and in particular a multi-stakeholder approach to the protection of the environment, the foundation of our lives. As mentioned previously, however, this will be possible only if ICTs are used for the common good, hence well governed. This means that ICTs are both an opportunity for global environmental governance actors and a pitfall. If not well managed, ICTs can increase insecurity, mistrust, and chaos among populations. Therefore, it is equally essential to develop good governance practices for the environment and digital technologies, which are open to all stakeholders, and which put the sustainable well-being of human populations at the center of decisions.

Notes

1 See Ess, Charles, 1996. The political computer: Democracy, CMC, and Habermas. In Ess, C. (ed.). *Philosophical perspectives on computer-mediated communication.* Albany: State University of New York Press, pp. 197–230.
2 See Ess, Charles, 2018. Democracy and the internet: A retrospective. *Javnost – The Public*, 25, pp. 1–2, 93–101. doi:10.1080/13183222.2017.1418820
3 Velazquez Gomar, J.O., 2014. Environmental policy integration among multilateral environmental agreements: The case of biodiversity. *International Environmental Agreements*, 16, p. 528. doi:10.1007/s10784-014-9263-4
4 Lovejoy, K., Saxton, G.D., 2012. Information, community, and action: How nonprofit organizations use social media. *Journal of Computer-Mediated Communication*, 17(3), p. 351. doi:10.1111/j.1083-6101.2012.01576.x

Index

Note: Page numbers in **bold** indicate a table on the corresponding page.